T0224078

Lecture Notes in Computer Science 10420

Commenced Publication in 1973
Founding and Former Series Editors:
Gerhard Goos, Juris Hartmanis, and Jan van Leeuwen

More information about this series at http://www.springer.com/series/8637

Abdelkader Hameurlain · Josef Küng
Roland Wagner · Sanjay Madria
Takahiro Hara (Eds.)

Transactions on Large-Scale Data- and Knowledge- Centered Systems XXXII

Special Issue on Big Data Analytics and Knowledge Discovery

Springer

Editors-in-Chief
Abdelkader Hameurlain
IRIT, Paul Sabatier University
Toulouse
France

Roland Wagner
FAW, University of Linz
Linz
Austria

Josef Küng
FAW, University of Linz
Linz
Austria

Guest Editors
Sanjay Madria
Missouri University of Science
  and Technology
Rolla, MO
USA

Takahiro Hara
Osaka University
Osaka
Japan

ISSN 0302-9743   ISSN 1611-3349   (electronic)
Lecture Notes in Computer Science
ISSN 1869-1994   ISSN 2510-4942   (electronic)
Transactions on Large-Scale Data- and Knowledge-Centered Systems
ISBN 978-3-662-55607-8   ISBN 978-3-662-55608-5   (eBook)
DOI 10.1007/978-3-662-55608-5

Library of Congress Control Number: 2017947502

Printed on acid-free paper

This Springer imprint is published by Springer Nature
The registered company is Springer-Verlag GmbH Germany
The registered company address is: Heidelberger Platz 3, 14197 Berlin, Germany

Preface

Big data are rapidly growing in all domains. Knowledge discovery using data analytics is important to several applications ranging from health care to manufacturing to smart city. This special issue is based on selected papers from the 2015 Conference on Big Data Analytics and Knowledge Discovery (DAWAK), which is a forum providing a platform for the exchange of ideas and experiences among theoreticians and practitioners who are involved in the design, management, and implementation of big data, analytics, and knowledge discovery solutions.

We selected only five papers from the 25 presented at the conferences, which are included in the DAWAK 2015 proceedings. These papers have been extended and peer reviewed and revised again.

Major credit for the quality of this guest issue goes to the authors who revised and resubmitted journal-quality papers and those reviewers who under tight deadlines completed the journal reviews. We would like to thank all the members of the DEXA committee for their support and help; in particular, we are very thankful to Gabriela Wagner for her non-stop support with this journal issue. Finally, we would like to thank the journal office bearers for allowing us to publish this special issue based on selected papers from the DAWAK 2015 conference.

May 2017
Sanjay Madria
Takahiro Hara

Organization

Editorial Board

Contents

Exact Detection of Information Leakage: Decidability and Complexity

Rada Chirkova[1(✉)] and Ting Yu[2]

[1] Computer Science Department,
North Carolina State University, Raleigh, NC, USA
chirkova@csc.ncsu.edu
[2] Qatar Computing Research Institute,
Hamad Bin Khalifa University, Doha, Qatar
tyu@qf.org.qa

Abstract. Elaborate security policies often require organizations to restrict user data access in a fine-grained manner, instead of traditional table- or column-level access control. Not surprisingly, managing fine-grained access control in software is rather challenging. In particular, if access is not configured carefully, information leakage may happen: Users may infer sensitive information through the data explicitly accessible to them.

In this paper we formalize this *information-leakage problem,* by modeling sensitive information as answers to "secret queries," and by modeling access-control rules as views. We focus on the scenario where sensitive information can be deterministically derived by adversaries. We review a natural data-exchange based inference model for detecting information leakage, and show its capabilities and limitation. We then introduce and formally study a new inference model, view-verified data exchange, that overcomes the limitation for the query language under consideration. Our formal study provides correctness and complexity results for the proposed inference model in the context of queries belonging to a frequent realistic query type and common types of integrity constraints on the data.

Keywords: Privacy and security in data-intensive systems · Information leakage · Data exchange

1 Introduction

In many data-intensive applications, access to sensitive information is regulated by *access-control policies* [8], which determine, for each user or class of users, the stored information that they are permitted to access. For instance, an employee working in the communications department of a company may be granted access to view in the employee database his salary, but not any other employee's salaries. At the same time, an employee of the human-resources department may be permitted to view the salaries of all the employees in the communications department. To enable management of information covered by heterogeneous policies

© Springer-Verlag GmbH Germany 2017
A. Hameurlain et al. (Eds.): TLDKS XXXII, LNCS 10420, pp. 1–23, 2017.
DOI: 10.1007/978-3-662-55608-5_1

such as in this example, it has become standard for data-management systems from major vendors to go beyond table-level or column-level access control, by providing means to support row-level or even cell-level access control. Representative products that provide such *fine-grained access control* include Oracle's Virtual Private Database [26] and DB2's label-based access control [10].

Unlike table-level or column-level access control, fine-grained access control is often content based. That is, whether a user is allowed access to some data depends on the content of the data, and often also on the properties of the user, as illustrated in the above example. Content-based access control is a powerful mechanism that offers flexibility to express nontrivial constraints on user data access, which is often a must for the enforcement of privacy and corporate policies. However, correct configuration of fine-grained access control is a challenging task in data-intensive systems. For instance, many database systems offer a programming-style interface for specifying access-control rules. In this clearly error-prone process, a simple programming error may result in serious data-access violations. Specifically, even if the access-control rules for each user are implemented correctly, sensitive information might still be derived by colluding malicious users from the data that they are permitted to access. Consider an illustration that introduces our running example.

Example 1. Suppose a database relation Emp stores the name, department, and salary information about employees of a company. Sensitive database information that should not be disclosed to unauthorized users could include the relationship between the employee names and salaries in the Emp relation. Suppose user X is permitted to access the name-department relationships in Emp, via a projection of that relation on the respective columns. Further, suppose another user, Y, is granted access to the department and salary columns of Emp in a similar way.

Consider an instance of the Emp relation on which the user X sees exactly one relationship between employee names and departments, namely the pair (JohnDoe, Sales). Suppose the user Y can access, from the same instance of Emp, also a single relationship pair, (Sales, $50000), and no other information. If X and Y collude, it is straightforward for them to infer from these two tuples that employee JohnDoe's salary must be $50000 in the Emp relation. Thus, the sensitive name-salary relationship can be obtained in this case by unauthorized users just by comparing the data that they are permitted to see.

This example illustrates how content-based access control may be exploited by malicious users to derive sensitive information. The derivation in the example is *deterministic*. That is, the presence of the sensitive relationship (JohnDoe, $50000) in the back-end database is guaranteed by the information that these users are authorized to access. We call the problem of discovering whether it is possible to make such deterministic derivations of sensitive information the *information-leakage problem.* Our specific focus is as follows: Given a set of access-control rules for a database, and given the current database status, will a user or set of colluding users be able to deterministically derive some secret information from their explicitly accessible data, and possibly from their knowledge of integrity constraints that must hold on the database? In this paper, we formalize and study this information-leakage problem in the context of fine-grained access control on relational data, with secret queries and access views belonging to a frequent realistic query type of SQL select-project-join *(SPJ)* queries with equality.

Our specific contributions are as follows:

- We formalize the information-leakage problem for access control on relational data, for a single secret query and one or more access-control views, possibly in presence of integrity constraints on the back-end database.
- We perform a theoretical study of this problem in the setting where the secret query and each access-control policy is a SQL select-project-join (SPJ) query with equalities. As such queries are also called *conjunctive (CQ) queries* [12], we will refer to this setting as *the CQ setting*. Specifically, we develop an algorithm for information-leak disclosure in the CQ setting using a common type of integrity constraints ("weakly acyclic dependencies" [18]), and prove its correctness and complexity.

In the original conference version [4] of this paper, the presentation focused on discussions, illustrations, and on outlining the results at the intuitive level, and treated the special case of the problem where the set of the given integrity constraints is empty. In this current paper, we present a full formalization of the problem, add a formal treatment, results, and proofs for the more general case potentially involving integrity constraints that commonly occur in practice, and introduce and prove new results concerning the correctness, validity, and complexity of the proposed view-verified data-exchange approach for this more general case.

The remainder of the paper is organized as follows. Section 2 formalizes the problem of information-leak disclosure. Sections 3 and 4 discuss two approaches to solving the problem: a natural approach based on data exchange (Sect. 3) and a more elaborate version, view-verified data exchange, with completeness guarantees (Sect. 4). Section 5 presents the formal correctness, validity, and complexity results for view-verified data exchange. Section 6 discusses related work, and Sect. 7 concludes.

2   The Problem Statement

In this section we formalize the problem of information-leak disclosure. As an overview, the problem is as follows: Given a set of answers, MV, to *access-control queries* $\mathcal{V}$ (which we call *views* in the rest of the paper), on an (unavailable) relational data set of interest I, and given another query Q, which we call *the secret query* in the sense that its answers must not be disclosed to unauthorized users: Which of the answer tuples to Q on I are "deterministically assured" by the contents of MV? That is, which tuples $\bar{t}$ must necessarily be present in the answer to Q on I, based on the information in $\mathcal{V}$, in MV, and (optionally) in the set Σ of *integrity constraints* that hold on I? We say that there is *information leakage* of Q via MV and $\mathcal{V}$ if and only if at least one such tuple $\bar{t}$ exists. Note that the data set I is not available for making the determination.

Formally, suppose we are given a relational schema $\mathbf{P}$, set of dependencies Σ on $\mathbf{P}$, set $\mathcal{V}$ of views on $\mathbf{P}$, query Q over $\mathbf{P}$, and a (Σ-valid) set MV of view answers for $\mathcal{V}$. (A set MV is Σ-*valid for* $\mathcal{V}$ whenever there exists a data set I that satisfies Σ and produces exactly MV as the set of answers to $\mathcal{V}$. We use the notation "$\mathcal{V} \Rightarrow_{I,\Sigma} MV$" to refer to such data sets I.) Then we call the tuple $\mathcal{D}_s = (\mathbf{P}, \Sigma, \mathcal{V}, Q, MV)$ a *(valid) instance of the information-leak problem*. We also distinguish within $\mathcal{D}_s$ the part $\mathcal{M}\Sigma = (\mathbf{P}, \Sigma, \mathcal{V}, MV)$, which we call the *(valid) materialized-view setting for* $\mathbf{P}$, Σ, $\mathcal{V}$, and MV.

We say that a tuple $\bar{t}$ is a *potential information leak* in instance $\mathcal{D}_s$ if for all I such that $\mathcal{V} \Rightarrow_{I,\Sigma} MV$ we have $\bar{t} \in Q(I)$. Further, we say that *there is a disclosure of an information leak* in $\mathcal{D}_s$ iff there exists a tuple $\bar{t}$, such that one can show deterministically, using only the information in $\mathcal{D}_s$, that $\bar{t}$ is a potential information leak in $\mathcal{D}_s$. In this case, such a $\bar{t}$ is called a *witness* of the disclosure.

The intuition here is as follows. Suppose attackers are examining the available set of view answers MV, trying to find answers to the secret query Q on the back-end data set, call it I, of schema $\mathbf{P}$ such that I has actually given rise to MV. (That is, I is an actual instance of the organizational data, such that $\mathcal{V} \Rightarrow_{I,\Sigma} MV$ and such that the set MV was obtained by applying to I the queries for $\mathcal{V}$.) However, assuming that the attackers have access only to the information in $\mathcal{D}_s$, they have no way of determining which data set J, with the property $\mathcal{V} \Rightarrow_{J,\Sigma} MV$, is that actual data set I behind the set MV. Thus, their best bet is to find all the tuples $\bar{t}$ that are in the answer to Q on *all* data sets J as above.

Definition 1 *(Problem of information-leak disclosure). Given an instance $\mathcal{D}_s$ of the information-leak problem, the* problem of information-leak disclosure for $\mathcal{D}_s$ *is to determine whether there is a potential information leak in $\mathcal{D}_s$.*

Note that in Definition 1, we do not require that the instance $\mathcal{D}_s$ of the information-leak problem be valid. The reason is, in Sect. 5 of this paper we will develop an approach to determining validity of instances of the information-leak problem, for the case where the input instances involve CQ queries and "weakly acyclic" [18] dependencies.[1] When an instance is not valid, the answer to the problem of information-leak disclosure is trivially negative.

The following example, which formalizes our running example from Sect. 1, provides an illustration of the definitions.

Example 2. Suppose a relation Emp stores information about employees of a company. The attributes of Emp are Name, Dept, and Salary, with self-explanatory names. A "secret query" Q (in SQL) asks for the salaries of all the employees:

---

[1] Weakly acyclic dependencies [7] are sets of tuple- and equality-generating integrity constraints (*tgds and egds,* respectively) that commonly occur in practice and have nice formal properties.

(Q): SELECT DISTINCT Name, Salary FROM Emp;

Consider two views, V and W, that are defined for some class(es) of users, in SQL on the schema **P**. The view V returns the department name for each employee, and the view W returns the salaries in each department:

(V): DEFINE VIEW V(Name, Dept) AS SELECT DISTINCT Name, Dept FROM Emp;
(W): DEFINE VIEW W(Dept, Salary) AS SELECT DISTINCT Dept, Salary FROM Emp;

Suppose some user(s) are authorized to see the answers to V and W. At some point the user(s) can see the following set MV of answers to these views:

$MV = \{$ V(JohnDoe, Sales), W(Sales, \$50000) $\}$.

Consider a tuple $\bar{t} =$ (JohnDoe, \$50000). Assume that these users are interested in finding out whether $\bar{t}$ is in the answer to the query Q on the "back-end" data set, I. Using the approach introduced in this paper, we can show that the relation Emp in such a data set I is uniquely determined by V, W, and MV:

$$\text{Emp } in \ I \ is \ \{\text{Emp}(\text{JohnDoe}, \text{Sales}, \$50000)\}. \tag{1}$$

(Note that the existence of this relation Emp satisfying all the input requirements means that the given instance of the information-leak problem is valid.)

The only answer to Q on this data set I is the above tuple $\bar{t}$. Thus, the presence of $\bar{t}$ in the answer to Q on the data set of interest is deterministically assured by the information that these users are authorized to access.

3 Using Data Exchange for Information-Leak Disclosure

We address the problem of information-leak disclosure by developing an approach that outputs the set of disclosed information leaks for each problem instance $\mathcal{D}_s$. Clearly, only those approaches that are sound and complete algorithms (for certain classes of inputs) can be used to solve correctly the problem stated in Definition 1 of Sect. 2. Here, *completeness* means that for each potential information-leak tuple $\bar{t}$ for a valid input $\mathcal{D}_s$, the given approach outputs $\bar{t}$. Further, *soundness* means that for each $\bar{t}$ that the approach outputs for such a $\mathcal{D}_s$, the tuple $\bar{t}$ is a potential information leak for $\mathcal{D}_s$.

We now begin the development of a sound and complete approach, by showing how data exchange [7] can be used for information-leak disclosure. The ideas in this section parallel those of [11,27]. These ideas are not a contribution of this paper, as they are sound but not complete in our context, please see Sect. 3.3. The main contribution of this paper is an approach that builds on but does not stop at data exchange; we present that approach in Sect. 4. The approach of Sect. 4 is sound and complete for information-leak disclosure for CQ access-control policies and secret queries, both in the absence and in presence of (weakly acyclic) integrity constraints, please see Sect. 5. We now begin laying down the framework for that approach.

3.1   Reviewing Data Exchange

A *data-exchange setting* $\mathcal{M}$ is a triple $(\mathbf{S}, \mathbf{T}, \Sigma_{st} \cup \Sigma_t)$, where $\mathbf{S}$ and $\mathbf{T}$ are disjoint schemas, Σ_{st} is a finite set of integrity constraints applying from $\mathbf{S}$ to $\mathbf{T}$,[2] and Σ_t is a finite set of integrity constraints over $\mathbf{T}$. Instances of $\mathbf{S}$ are called *source* data sets and are always ground data sets (i.e., never include "null values," also called "nulls"). Instances of $\mathbf{T}$ are *target* data sets.

Given a source data set I, we say that a target data set J is a *solution for I (under $\mathcal{M}$)* if (I, J) satisfies $\Sigma_{st} \cup \Sigma_t$, denoted "$(I, J) \models \Sigma_{st} \cup \Sigma_t$." Solutions are not always unique; *universal solutions* are, intuitively, "the most general" possible solutions. Constructing a universal solution for source data set I can be done by chasing [18] I with the dependencies (typically egds and tgds [2]) in $\Sigma_{st} \cup \Sigma_t$. The chase may not terminate or may fail; in the latter case, no solution exists [18]. If the chase does not fail and terminates, then the resulting target data set is guaranteed to be a universal solution for I.

While checking for existence of solutions is undecidable [7], a positive result is known for the case where Σ_t is a set of weakly acyclic [18] dependencies.

Theorem 1 [18]. *Let $\mathcal{M} = (\mathbf{S}, \mathbf{T}, \Sigma_{st} \cup \Sigma_t)$ be a fixed data-exchange setting, with Σ_t a set of weakly acyclic egds and tgds. Then there is a polynomial-time algorithm, which decides for every source data set I whether a solution for I exists. Whenever a solution for I exists, the algorithm computes a universal solution in polynomial time.*

The universal solution of Theorem 1, called the *canonical* universal solution [18], is the result of the chase.

Query answering: Assume that a user poses a query Q over the target schema $\mathbf{T}$, and I is a given source data set. Then the usual semantics for the query answering is that of *certain answers* [7], which intuitively are tuples that occur in the answer to Q on all possible solutions for I in $\mathcal{M}$:

$$certain_{\mathcal{M}}(Q, I) = \bigcap \{ \, Q(J) \mid J \text{ is a solution for } I \, \}.$$

Computing certain answers for arbitrary first-order queries is an undecidable problem. For unions of CQ queries *(UCQ queries)*, which is a query language including all CQ queries, we have the following positive result:

Theorem 2 [18]. *Let $\mathcal{M} = (\mathbf{S}, \mathbf{T}, \Sigma_{st} \cup \Sigma_t)$ be a data-exchange setting with Σ_t a weakly acyclic set, and let Q be a UCQ query. Then the problem of computing certain answers for Q under $\mathcal{M}$ can be solved in polynomial time.*

To compute the certain answers to a UCQ query Q with respect to a source data set I, we check whether a solution for I exists. If so, compute any universal solution J for I, and compute the set $Q_{\downarrow}(J)$ of all the tuples in $Q(J)$ that do not contain nulls. It can be shown that $Q_{\downarrow}(J) = certain_{\mathcal{M}}(Q, I)$.

---

[2] The intuition is that tuple patterns occuring over $\mathbf{S}$ constrain tuple patterns over $\mathbf{T}$.

3.2   Data Exchange in Information-Leak Disclosure

We now show how one can reformulate an instance of the information-leak disclosure problem of Definition 1 into an instance of the problem of finding certain query answers in data-exchange settings. We begin by formalizing the certain-query-answer problem in a materialized-view setting.

Let $\mathcal{M\Sigma}$ be a materialized-view setting $(\mathbf{P}, \Sigma, \mathcal{V}, MV)$, and let Q be a query over $\mathbf{P}$. We define the *set of certain answers of Q w.r.t. the setting $\mathcal{M\Sigma}$* as

$$certain_{\mathcal{M\Sigma}}(Q) = \bigcap \{Q(I) \text{ where } I \text{ s.t.} \mathcal{V} \Rightarrow_{I,\Sigma} MV \text{ in } \mathcal{M\Sigma}\}.$$

That is, the set of certain answers of a query w.r.t. a setting is understood, as usual, as the set of all tuples that are in the answer to the query on all the instances relevant to the setting. (Cf. [1] for the case $\Sigma = \emptyset$.)

Definition 2 *(Certain-query-answer problem in a materialized-view setting). Given a setting $\mathcal{M\Sigma} = (\mathbf{P}, \Sigma, \mathcal{V}, MV)$, a k-ary $(k \geq 0)$ query Q over the schema $\mathbf{P}$ in $\mathcal{M\Sigma}$, and a ground k-tuple $\bar{t}$. Then the* certain-query-answer problem for Q and $\bar{t}$ in $\mathcal{M\Sigma}$ *is to determine whether $\bar{t} \in certain_{\mathcal{M\Sigma}}(Q)$.*

It is easy to show that a tuple $\bar{t}$ can be a certain answer to a query Q in a setting $\mathcal{M\Sigma}$ only if all the values in $\bar{t}$ are in $consts(\mathcal{M\Sigma})$, which denotes the set of constants occurring in $\mathcal{M\Sigma}$. (For a given materialized-view setting $\mathcal{M\Sigma} = (\mathbf{P}, \Sigma, \mathcal{V}, MV)$, we define $consts(\mathcal{M\Sigma})$ as the union of the set $adom(MV)$ of all the constant values used in the relations in MV with the set of all the constants used in the definitions of the views $\mathcal{V}$.) By this observation, in Definition 2 we can restrict our consideration to the tuples $\bar{t}$ with this property.

Suppose that we are given a valid CQ materialized-view setting $\mathcal{M\Sigma} = (\mathbf{P}, \Sigma, \mathcal{V}, MV)$ and a CQ query Q of arity $k \geq 0$. Recall that $\mathcal{M\Sigma}$ and Q together comprise an instance of the information-leak problem. By the above definitions, in the problem of information-leak disclosure we are interested in finding all (and only) k-ary tuples $\bar{t}$ of elements of $consts(\mathcal{M\Sigma})$, such that for all the instances I satisfying $\mathcal{V} \Rightarrow_{I,\Sigma} MV$, we have $\bar{t} \in Q(I)$.

We are now ready to show how a straightforward reformulation of $\mathcal{M\Sigma}$ and Q turns the above problem into an instance of the problem of computing certain answers in data exchange. We first construct a set Σ_{st} of tuple-generating dependencies *(tgds)*, as follows. For a view V in the set of views $\mathcal{V}$ in $\mathcal{M\Sigma}$, consider the query $V(\bar{X}) \leftarrow body_{(V)}(\bar{X}, \bar{Y})$ for V. (As $\mathcal{M\Sigma}$ is a CQ instance, the query for each V in $\mathcal{V}$ is a CQ query.) We associate with this $V \in \mathcal{V}$ the tgd $\sigma_V : V(\bar{X}) \rightarrow \exists \bar{Y} \ body_{(V)}(\bar{X}, \bar{Y})$. We then define the set Σ_{st} to be the set of tgds σ_V for all $V \in \mathcal{V}$. Then $\mathcal{M\Sigma}$ can be reformulated into a data-exchange setting

$$\mathcal{S}^{(de)}(\mathcal{M\Sigma}) = (\ \mathcal{V}, \mathbf{P}, \Sigma_{st} \cup \Sigma \ ),$$

with a source instance MV and a query Q on the target schema $\mathbf{P}$. We call the triple $(\mathcal{S}^{(de)}(\mathcal{M\Sigma}), MV, Q)$ the *associated data-exchange instance for $\mathcal{M\Sigma}$ and Q.*

The observation of the following Proposition 1 is immediate from Definition 2 and from the definitions in Sect. 2. For this observation and for the exposition starting in the next subsection, we recall the notions of the "closed-world assumption" *(CWA)* and of the "open-world assumption" *(OWA)*, please see [1]. Intuitively, for a base instance I, consider the answer tuples generated by the given view definitions on I. Then, informally, I is relevant to the given instance MV of materialized views under CWA iff these answer tuples together comprise exactly MV. In contrast, OWA permits MV to not contain all such tuples. CWA is the main focus in our context, because in the problem of information-leak disclosure, potential attackers are interested in obtaining answers to secret queries only from those base (back-end) databases whose answers to the views are exactly those in MV. Example 2 of Sect. 2 illustrates the CWA perspective on the data set I; if any other tuples were added to the relation Emp in I in that example, then the example would shift to the OWA perspective, as the resulting sets of view answers could become proper supersets of the given set MV. (Example 3 in Sect. 3.3 will elaborate further on the OWA-CWA distinction in this context.)

Proposition 1. *Given a valid CQ materialized-view setting $\mathcal{M}\Sigma$ and a CQ query Q, with their associated data-exchange instance $(\mathcal{S}^{(de)}(\mathcal{M}\Sigma), MV, Q)$. Then for each tuple $\bar{t}$ that is a certain answer of Q with respect to MV under the data-exchange setting $\mathcal{S}^{(de)}(\mathcal{M}\Sigma)$, we have that $\bar{t}$ is a certain answer to the query Q w.r.t. $\mathcal{M}\Sigma$ under CWA.*

For valid CQ weakly acyclic settings $\mathcal{M}\Sigma$ and CQ queries Q, we introduce the following *data-exchange approach to finding certain query answers w.r.t. a materialized-view setting.* First, we compute the canonical universal solution, $J_{de}^{\mathcal{M}\Sigma}$, for the source instance MV in the data-exchange setting $\mathcal{S}^{(de)}(\mathcal{M}\Sigma)$. If $J_{de}^{\mathcal{M}\Sigma}$ does not exist, then we output the empty set of answers. Otherwise we output, as a set of certain answers to Q w.r.t. $\mathcal{M}\Sigma$, the set of all those tuples in $Q(J_{de}^{\mathcal{M}\Sigma})$ that do not contain nulls. The following result is immediate from Theorem 2 due to [18]:

Proposition 2. *Given a valid CQ weakly acyclic setting $\mathcal{M}\Sigma$ and a CQ query Q. Suppose that everything in $\mathcal{M}\Sigma$ is fixed except for MV, and suppose that Q is not fixed. Then the data-exchange approach to finding certain query answers w.r.t. a materialized-view setting always terminates when applied to $\mathcal{M}\Sigma$ and Q. Further, the approach applied to $\mathcal{M}\Sigma$ and Q constructs the respective instance $J_{de}^{\mathcal{M}\Sigma}$ in polynomial time, and returns each certain-answer tuple in polynomial time.*

(The assumption of Proposition 2 that everything in $\mathcal{M}\Sigma$ is fixed except for MV, and that Q is not fixed, comes from the informaton-security setting of [29]. This complexity assumption is appropriate both in the context of this section and in the context of our proposed view-verified data-exchange approach, as discussed in Sect. 5.5.)

By Proposition 2, the data-exchange approach to finding certain query answers w.r.t. a materialized-view setting is an algorithm when applied to valid CQ weakly acyclic settings and CQ queries; we refer to it as the *data-exchange algorithm*. By Proposition 1, the data-exchange algorithm is sound:

Theorem 3. *The data-exchange algorithm is sound for valid weakly acyclic CQ inputs.*

3.3   The Data-Exchange Approach Is Not Complete

The reason we are to introduce the view-verified data-exchange algorithm of Sect. 4 is that, not surprisingly, the data-exchange algorithm of Sect. 3.2 is not complete under CWA for CQ queries, either for CQ settings $\mathcal{M}\Sigma$ with $\Sigma = \emptyset$, or for CQ weakly acyclic settings with $\Sigma \neq \emptyset$. In the remainder of this section, we discuss a feature of the data-exchange algorithm that prevents us from using it as a complete algorithm for this class of input instances under CWA. In Sect. 4 we will eliminate this feature of the data-exchange algorithm, in a modification that will give us a sound and complete algorithm for finding all and only the certain-answer tuples for CQ weakly acyclic settings and CQ queries under CWA.

We now prove that the data-exchange algorithm is not complete for CQ instances $\mathcal{M}\Sigma$ with $\Sigma = \emptyset$.

Example 3. We recall the CQ query Q and the CQ views V and W of Example 2 (Sect. 2). We rewrite the definitions here into Datalog, with extensional relation symbol E standing for the relation Emp of Example 2:

$$Q(X, Z) \leftarrow E(X, Y, Z).$$
$$V(X, Y) \leftarrow E(X, Y, Z).$$
$$W(Y, Z) \leftarrow E(X, Y, Z).$$

Suppose that c, d, and f stand for 'JohnDoe', 'Sales', and '\$50000', respectively. By applying this "translation of constants" to the instance MV of Example 2, we obtain an instance $MV = \{V(c, d), W(d, f)\}$. In the same notation, the tuple $\bar{t}$ of Example 2 is recast as (c, f).

Consider the materialized-view setting $\mathcal{M}\Sigma = (\{E\}, \emptyset, \{V, W\}, MV)$, with all the elements as defined above. By definition, $\mathcal{M}\Sigma$ is a CQ weakly acyclic setting. ($\mathcal{M}\Sigma$ is also valid, by the existence of the instance $\{(c, d, f)\}$ of schema $\{E\}$ as discussed in Example 2.) The data-exchange algorithm of Sect. 3.2 applied to $\mathcal{M}\Sigma$ and Q yields the following canonical universal solution, $J_{de}^{\mathcal{M}\Sigma}$, for the source instance MV in the data-exchange setting $\mathcal{S}^{(de)}(\mathcal{M}\Sigma)$:

$$J_{de}^{\mathcal{M}\Sigma} = \{ \; E(c, d, \perp_1), \; E(\perp_2, d, f) \; \}.$$

The first tuple in $J_{de}^{\mathcal{M}\Sigma}$ is due to the tuple $V(c, d)$ in MV, and the second tuple is due to $W(d, f)$ in MV. (The symbols of the form $\perp_i$, here and in the remainder of the paper, stand for null values.)

It is easy to verify that each of the two answers to the query Q on the instance $J_{de}^{\mathcal{M}\Sigma}$ has nulls and thus cannot qualify as a certain answer to Q w.r.t. $\mathcal{M}\Sigma$.

The reason is, the instance $J_{de}^{\mathcal{M}\Sigma}$ is an OWA base instance in the context of Example 2. As such, $J_{de}^{\mathcal{M}\Sigma}$ can express both the instance I of Example 2 (by the substitution of f for $\bot_1$ and of c for $\bot_2$), as well as other instances compatible with (but possibly also extending) the given set MV. In contrast, the instance I of Example 2 is a CWA base instance for the same problem input, as the answers to the views V and W on that instance comprise exactly the given set MV and nothing else.

When given as inputs the setting $\mathcal{M}\Sigma$ and query Q of Example 3, the data-exchange algorithm of Sect. 3.2 outputs the empty set of candidate-answer tuples. It is easy to verify that the tuple $\bar{t}$ of Example 2 should actually be in the answer (i.e., in the set of candidate-answer tuples) for the problem input of Example 3. As Q is a CQ query and $\mathcal{M}\Sigma$ is CQ weakly acyclic (with $\Sigma = \emptyset$), we conclude that the data-exchange algorithm is incomplete when applied to CQ queries and CQ weakly acyclic settings with $\Sigma = \emptyset$. Further, we can exhibit examples showing that the data-exchange algorithm is also incomplete when applied to (CQ queries and) CQ weakly acyclic materialized-view settings with $\Sigma \neq \emptyset$.

Why is the data-exchange algorithm not complete when applied to (CQ queries and) CQ weakly acyclic materialized-view settings? Intuitively, the problem is that its canonical universal solution $J_{de}^{\mathcal{M}\Sigma}$ "may cover too many target instances" (i.e., $J_{de}^{\mathcal{M}\Sigma}$ is an OWA rather than CWA solution). Let us evaluate the queries for the views V and W of Example 3 over the solution $J_{de}^{\mathcal{M}\Sigma}$ of that example. We obtain that the answer to the view V on $J_{de}^{\mathcal{M}\Sigma}$ is $\{V(c,d),\ V(\bot_2,d)\}$. Similarly, the answer to W on $J_{de}^{\mathcal{M}\Sigma}$ is $\{W(d,\bot_1),\ W(d,f)\}$. Thus, if we replace $\bot_1$ in $J_{de}^{\mathcal{M}\Sigma}$ by any constant except f, or replace $\bot_2$ by any constant except c, then any ground instance obtained from $J_{de}^{\mathcal{M}\Sigma}$ using these replacements would "generate too many tuples" (as compared with MV) in the answer to either V or W.

We now generalize over this observation. Fix a valid CQ weakly acyclic instance $\mathcal{M}\Sigma$, and consider the canonical universal solution (if one exists) $J_{de}^{\mathcal{M}\Sigma}$ generated by the data-exchange algorithm with $\mathcal{M}\Sigma$ as input. (In the remainder of this paper, we will refer to $J_{de}^{\mathcal{M}\Sigma}$ as *the canonical data-exchange solution for* $\mathcal{M}\Sigma$.) By definition of $J_{de}^{\mathcal{M}\Sigma}$, for each $V \in \mathcal{V}$, the answer to the query for V on $J_{de}^{\mathcal{M}\Sigma}$ is a superset of the relation $MV[V]$. Suppose that the answer on $J_{de}^{\mathcal{M}\Sigma}$ to at least one view $V \in \mathcal{V}$ is not a subset of $MV[V]$, as in, e.g., Example 3 just discussed. Then $J_{de}^{\mathcal{M}\Sigma}$, as a template for instances of schema $\mathbf{P}$, describes not only instances that "generate" exactly the set MV in $\mathcal{M}\Sigma$, but also those instances that generate proper supersets of MV. The latter instances are not of interest to us. (Recall that we take the CWA viewpoint, and thus are interested only in the instances I of schema $\mathbf{P}$ such that $\mathcal{V} \Rightarrow_{I,\Sigma} MV$.) As a result, when the data-exchange algorithm uses $J_{de}^{\mathcal{M}\Sigma}$ to obtain certain answers to the input query Q, it can easily miss those certain answers that characterize only those instances of interest to us.

4   View-Verified Data Exchange

We now build on the intuition and results of Sect. 3, to introduce an approach that is provably sound and complete for information-leak disclosure for conjunctive access-control policies and secret queries.

The problem with the natural data-exchange approach, as introduced in [11,27], is that its canonical universal solution, when turned into a ground instance, may produce a proper superset of the given set of view answers MV. (See Sect. 3.3 in this current paper.) That is, the canonical data-exchange solution does not necessarily describe ground solutions for $\mathcal{M}\Sigma$ "tightly enough." (Recall that we take the CWA viewpoint, and thus are interested only in the instances I of schema $\mathbf{P}$ such that $\mathcal{V} \Rightarrow_{I,\Sigma} MV$. At the same time, the canonical data-exchange solution describes not only these "CWA" instances, but also those that are relevant to the inputs under OWA.)

The approach that we introduce in this section builds on data exchange, by "tightening" its universal solutions using $consts(\mathcal{M}\Sigma)$. This approach, which we call *view-verified data exchange,* solves correctly the problem of finding all the candidate-answer tuples w.r.t. a CQ query and a valid CQ weakly acyclic materialized-view setting. We also use the approach of this section to solve the problem of deciding whether a given materialized-view setting is valid, please see Sect. 5.3.

4.1   Chase with MV-Induced Dependencies

In Sect. 4.2 we will define view-verified data exchange for CQ weakly acyclic input instances. (Throughout this section, we use the term "CQ weakly acyclic input instance" to refer to a pair $(\mathcal{M}\Sigma, Q)$, where $\mathcal{M}\Sigma$ is a CQ weakly acyclic materialized-view setting, and Q is a CQ query over the schema $\mathbf{P}$ in $\mathcal{M}\Sigma$.) Given a $\mathcal{M}\Sigma$ with set of views $\mathcal{V}$ and set of view answers MV, the idea of the approach is to force the canonical data-exchange solution $J_{de}^{\mathcal{M}\Sigma}$ for $\mathcal{M}\Sigma$ to generate only the relations in MV as answers to the queries for $\mathcal{V}$. (By definition of $J_{de}^{\mathcal{M}\Sigma}$, the answer on $J_{de}^{\mathcal{M}\Sigma}$ to the query for each $V \in \mathcal{V}$ is always a superset of the relation $MV[V]$.) We achieve this goal by chasing $J_{de}^{\mathcal{M}\Sigma}$ using "MV-induced" dependencies. Intuitively, applying MV-induced dependencies to the instance $J_{de}^{\mathcal{M}\Sigma}$ forces some nulls in $J_{de}^{\mathcal{M}\Sigma}$ to become constants in $consts(\mathcal{M}\Sigma)$. As a result of such a chase step, we obtain that for at least one view $V \in \mathcal{V}$, some formerly non-ground tuples in the answer to V on the instance become ground tuples in $MV[V]$.

We now formally define MV-induced dependencies. Let $V(\bar{X}) \leftarrow \phi(\bar{X}, \bar{Y})$ be a CQ query of arity $k_V \geq 0$, and MV be a ground instance of a schema that includes the k_V-ary relation symbol V. First, in case where $MV[V] = \emptyset$, we define the *MV-induced implication constraint (MV-induced ic) ι_V for V* as

$$\iota_V : \ \phi(\bar{X}, \bar{Y}) \rightarrow \ false \, . \tag{2}$$

(Each MV-induced ic is an implication constraint, i.e., a Horn rule with the empty head. See [28] for the discussion and references on implication constraints.)

Second, in case where $k_V \geq 1$, suppose $MV[V] = \{\bar{t}_1, \bar{t}_2, \ldots, \bar{t}_{m_V}\}$, with $m_V \geq 1$. Then we define the *MV-induced generalized equality-generating dependency (MV-induced ged)* τ_V for V as

$$\tau_V : \phi(\bar{X}, \bar{Y}) \rightarrow \vee_{i=1}^{m_V} (\bar{X} = \bar{t}_i). \tag{3}$$

Here, $\bar{X} = [S_1, \ldots, S_{k_V}]$ is the head vector of the query for V, with $S_j \in \text{CONST} \cup \text{QVAR}$ for $j \in [1, k_V]$. For each $i \in [1, m_V]$ and for the ground tuple $\bar{t}_i = (c_{i1}, \ldots, c_{ik_V}) \in MV[V]$, we abbreviate by $\bar{X} = \bar{t}_i$ the conjunction $\wedge_{j=1}^{k_V} (S_j = c_{ij})$. *MV*-induced geds are a straightforward generalization of disjunctive equality-generating dependencies of [16,18].

We now define chase of instances with *MV*-induced dependencies. Consider first *MV*-induced implication constraints. Given an instance K of schema $\mathbf{P}$ and an *MV*-induced ic ι_V as in Eq. (2), suppose there exists a homomorphism h from the antecedent $\phi(\bar{X}, \bar{Y})$ of ι_V to K. The intuition here is that we want to make sure that K does not "generate" any tuples in the relation $MV[V]$; however, by the existence of h, the instance K does generate at least one such tuple. We then say that *chase with ι_V (and h) fails on the instance K and produces the set $\{\epsilon\}$*, with ϵ denoting the empty instance.

Now let τ as in Eq. (3) be an *MV*-induced generalized egd for a $V \in \mathcal{V}$. The intuition here is that K must "generate" *only* the tuples in the relation $MV[V]$; we make this happen by assigning nulls in K to constants in $MV[V]$. (If such assignments are not possible, chase with τ fails on K.) Example 4, the running example for this section, provides a detailed illustration.

Our definition of the chase step with τ as in Eq. (3) is a straightforward extension of the definition of [18] for their disjunctive egds, as follows. Consider the consequent of τ, of the form $\vee_{i=1}^{m_V} (\bar{X} = \bar{t}_i)$. Recall that for each $i \in [1, m_V]$, the expression $\bar{X} = \bar{t}_i$ is of the form $\wedge_{j=1}^{k_V} (S_j = c_{ij})$. Denote by $\tau^{(1)}, \ldots, \tau^{(m_V)}$ the following m_V dependencies obtained from τ: $(\phi(\bar{X}, \bar{Y}) \rightarrow \bar{X} = \bar{t}_1), \ldots, (\phi(\bar{X}, \bar{Y}) \rightarrow \bar{X} = \bar{t}_{m_V})$, and call them *the dependencies associated with τ*. For each $i \in [1, m_V]$, $\tau^{(i)}$ is an embedded dependency that can be equivalently represented by k_V egds $\tau^{(i,1)}, \ldots, \tau^{(i,k_V)}$. Here, for each $j \in [1, k_V]$, the egd $\tau^{(i,j)}$ is $\phi(\bar{X}, \bar{Y}) \rightarrow S_j = c_{ij}$.

Given a τ as in Eq. (3) and an instance K of schema $\mathbf{P}$, suppose that there exists a homomorphism h from $\phi(\bar{X}, \bar{Y})$ to K such that $\wedge_{j=1}^{k_V} (h(S_j) = h(c_{ij}))$ is not a tautology for any $i \in [1, m_V]$. Then we say that τ *is applicable to K with the homomorphism h*. It is easy to see that it is also the case that each of $\tau^{(1)}, \ldots, \tau^{(m_V)}$ can be applied to K with h. That is, for each $i \in [1, m_V]$, the chase of K is applicable with at least one egd $\tau^{(i,j)}$ in the equivalent representation of $\tau^{(i)}$ as a set of egds. For each $i \in [1, m_V]$, let K_i be the result of applying all the egds $\tau^{(i,1)}, \ldots, \tau^{(i,k_V)}$ to K with h. Note that chase with $\tau^{(i,j)}$ and h can fail on K for some i and j. For each such i, we say that *chase with $\tau^{(i)}$ fails on K and produces the empty instance ϵ*.

Similarly to [18], we distinguish two cases:

- If the set $\{K_1, \ldots, K_{m_V}\}$ contains only empty instances, we say that *chase with τ (and h) fails on K and produces the set $\{\epsilon\}$*.

– Otherwise, let $\mathcal{K}^{(\tau)} = \{K_{i_1}, \ldots, K_{i_p}\}$ be the set of all nonempty elements of $\{K_1, \ldots, K_{m_V}\}$. We say that $\mathcal{K}^{(\tau)}$ *is the result of applying τ to K with h.*

Similarly to the approach of [18], in addition to chase steps with MV-induced dependencies we will also use chase steps with egds and tgds as in [18]. For the chase step of each type, we will use the set notation for uniformity: $K \Rightarrow^{\sigma,h} \mathcal{K}'$ denotes that a chase step with dependency σ and homomorphism h applied to instance K yields a set of instances $\mathcal{K}'$. Whenever chase with an egd fails on K, the set $\mathcal{K}'$ is the set $\{\epsilon\}$ by convention; in all other cases where σ is an egd or a tgd, the set $\mathcal{K}'$ is a singleton set. For σ of the form as in Eqs. (2) and (3), the set $\mathcal{K}'$ is in some cases $\{\epsilon\}$ as defined above.

Definition 3 *(MV-enhanced chase). Let Σ be a set of egds and tgds, let $\Sigma^{(MV)}$ be a set of MV-induced dependencies, and let K be an instance.*

– *A chase tree of K with $\Sigma \cup \Sigma^{(MV)}$ is a tree (finite or infinite) such that:*
   • *The root is K, and*
   • *For every node K_j in the tree, let $\mathcal{K}_j$ be the set of its children. Then there must exist a dependency σ in $\Sigma \cup \Sigma^{(MV)}$ and a homomorphism h such that $K_j \Rightarrow^{\sigma,h} \mathcal{K}_j$.*
– *A finite MV-enhanced chase of K with $\Sigma \cup \Sigma^{(MV)}$ is a finite chase tree T, such that for each leaf K_p of T, we have that either (a) K_p is ϵ, or (b) there is no dependency σ in $\Sigma \cup \Sigma^{(MV)}$ and no homomorphism h such that σ can be applied to K_p with h.*

Example 4. Consider Q as in Example 3 (Sect. 3.3) and $\mathcal{M}\Sigma = (\{E\}, \emptyset, \{V, W\}, MV)$, with all the elements except MV as in Example 3. For this current example, we define the set MV as

$$MV = \{V(c, d), V(g, d), W(d, f)\}.$$

By definition, $\mathcal{M}\Sigma$ paired with Q is a CQ instance with $\Sigma = \emptyset$. $\mathcal{M}\Sigma$ is also valid, as witnessed by the instance $\{E(c, d, f), E(g, d, f)\}$. The data-exchange approach of Sect. 3 yields the following canonical data-exchange solution $J_{de}^{\mathcal{M}\Sigma}$ for $\mathcal{M}\Sigma$:

$$J_{de}^{\mathcal{M}\Sigma} = \{E(c, d, \perp_1), E(g, d, \perp_2), E(\perp_3, d, f)\}.$$

The set of answers without nulls to the query Q on $J_{de}^{\mathcal{M}\Sigma}$ is empty. Thus, the data-exchange approach applied to $\mathcal{M}\Sigma$ discovers no certain answers to the query Q w.r.t. the setting $\mathcal{M}\Sigma$.

In applying the view-verified data-exchange approach to the input $(\mathcal{M}\Sigma, Q)$, we first construct the MV-induced generalized egds, τ_V and τ_W, one for each of the two views in $\mathcal{M}\Sigma$. (As MV has no empty relations, we do not need to construct MV-induced ics for $\mathcal{M}\Sigma$.)

$\tau_V : E(X, Y, Z) \rightarrow (X = c \wedge Y = d) \vee (X = g \wedge Y = d).$
$\tau_W : E(X, Y, Z) \rightarrow (Y = d \wedge Z = f).$

The two dependencies associated with τ_V are $\tau_V^{(1)} : E(X, Y, Z) \to (X = c \wedge Y = d)$ and $\tau_V^{(2)} : E(X, Y, Z) \to (X = g \wedge Y = d)$. Each of $\tau_V^{(1)}$ and $\tau_V^{(2)}$ can be equivalently represented by two egds. For instance, the egd representation for $\tau_V^{(1)}$ is via $\tau_V^{(1,1)} : E(X, Y, Z) \to X = c$ and $\tau_V^{(1,2)} : E(X, Y, Z) \to Y = d$. Similarly, there is one dependency $\tau_W^{(1)}$ $(= \tau_W)$ associated with τ_W; an equivalent representation of $\tau_W^{(1)}$ is via two egds.

Consider a homomorphism $h_V^{(1)} : \{X \to c, Y \to d, Z \to \perp_1\}$ from the antecedent $E(X, Y, Z)$ of τ_V to the instance $J_{de}^{\mathcal{M\Sigma}}$. As applying $h_V^{(1)}$ to the consequent of $\tau_V^{(1)}$ gives us the tautology $(c = c \wedge d = d)$, we conclude that τ_V is not applicable to $J_{de}^{\mathcal{M\Sigma}}$ with $h_V^{(1)}$.

Consider now the homomorphism $h_V^{(2)} : \{X \to \perp_3, Y \to d, Z \to f\}$ from the antecedent of τ_V to $J_{de}^{\mathcal{M\Sigma}}$. Applying $h_V^{(2)}$ to the consequent of τ_V gives us the expression $(\perp_3 = c \wedge d = d) \vee (\perp_3 = g \wedge d = d)$, which has no tautologies among its disjuncts. Thus, τ_V is applicable to $J_{de}^{\mathcal{M\Sigma}}$ with $h_V^{(2)}$. The chase step with τ_V and $h_V^{(2)}$ transforms $J_{de}^{\mathcal{M\Sigma}}$ into instances J_1 and J_2, as follows.

$J_1 = \{E(c, d, \perp_1), E(g, d, \perp_2), E(c, d, f)\}$.
$J_2 = \{E(c, d, \perp_1), E(g, d, \perp_2), E(g, d, f)\}$.

(J_1 results from assigning $\perp_3 := c$, and J_2 from $\perp_3 := g$.)

We then use the same procedure to apply τ_W to each of J_1 and J_2. In each case, the chase steps assign the value f to each of $\perp_1$ and $\perp_2$. As a result, the following instance $J_{vv}^{\mathcal{M\Sigma}}$ is obtained from each of J_1 and J_2:

$J_{vv}^{\mathcal{M\Sigma}} = \{ E(c, d, f), E(g, d, f) \}$.

4.2 Solving CQ Weakly Acyclic Instances

We now define the view-verified data-exchange approach to the problem of finding all certain answers to queries w.r.t. materialized-view-settings.

Let $\mathcal{M\Sigma} = (\mathbf{P}, \Sigma, \mathcal{V}, MV)$ be a CQ materialized-view setting. Then *the set $\Sigma^{(\mathcal{M\Sigma})}$ of MV-induced dependencies for $\mathcal{M\Sigma}$* is a set of up to $|\mathcal{V}|$ elements, as follows. For each $V \in \mathcal{V}$ such that $k_V \neq 0$ or $MV[V] \neq \{()\}$, $\Sigma^{(\mathcal{M\Sigma})}$ has one MV-induced implication constraint or one MV-induced generalized egd, by the rules as in Eqs. (2) and (3) in Sect. 4.1.[3]

For CQ weakly acyclic input instances $(\mathcal{M\Sigma}, Q)$ we introduce the following *view-verified data-exchange approach to finding certain query answers w.r.t. a materialized-view setting*. First, we compute (as in Sect. 3) the canonical universal solution $J_{de}^{\mathcal{M\Sigma}}$ for the source instance MV in the data-exchange setting $\mathcal{S}^{(de)}(\mathcal{M\Sigma})$. If $J_{de}^{\mathcal{M\Sigma}}$ does not exist, we stop and output the answer that $\mathcal{M\Sigma}$ is not valid. Otherwise we obtain a chase tree of $J_{de}^{\mathcal{M\Sigma}}$ with $\Sigma \cup \Sigma^{(\mathcal{M\Sigma})}$, where

---

[3] We omit from $\Sigma^{(\mathcal{M\Sigma})}$ the dependencies, of the form $\phi(\bar{X}, \bar{Y}) \to true$, for the case where $k_V = 0$ *and* $MV[V] \neq \emptyset$. By the results in this section, adding these dependencies to $\Sigma^{(\mathcal{M\Sigma})}$ would not change any chase results.

$\Sigma^{(\mathcal{M}\Sigma)}$ is the set of MV-induced dependencies for $\mathcal{M}\Sigma$. If the chase tree is finite, denote by $\mathcal{J}_{vv}^{\mathcal{M}\Sigma}$ the set of all the nonempty leaves of the tree. We call each $J \in \mathcal{J}_{vv}^{\mathcal{M}\Sigma}$ a *view-verified universal solution for* $\mathcal{M}\Sigma$. If $\mathcal{J}_{vv}^{\mathcal{M}\Sigma} = \emptyset$, then we stop and output the answer that $\mathcal{M}\Sigma$ is not valid. Otherwise, for each $J \in \mathcal{J}_{vv}^{\mathcal{M}\Sigma}$ we compute the set $Q_{\downarrow}(J)$ of all the tuples in $Q(J)$ that do not contain nulls. Finally, the output of the approach for the input $(\mathcal{M}\Sigma, Q)$ is the set

$$\bigcap_{J \in \mathcal{J}_{vv}^{\mathcal{M}\Sigma}} Q_{\downarrow}(J) . \tag{4}$$

The view-verified data-exchange approach to the problem of finding all certain answers to queries w.r.t. materialized-view-settings addresses the shortcoming of the data-exchange approach, see Sect. 3.3. Recall that the canonical universal solution $J_{de}^{\mathcal{M}\Sigma}$ of the latter approach might not cover "tightly enough" all the instances of interest to the attackers. In the view-verified approach, we address this problem, by using our extension of the chase to generate from $J_{de}^{\mathcal{M}\Sigma}$ a set $\mathcal{J}_{vv}^{\mathcal{M}\Sigma}$ of instances that are each "tighter" than $J_{de}^{\mathcal{M}\Sigma}$ in this sense.

In Sect. 5 we will show that the view-verified data-exchange approach is a sound and complete algorithm for the problem of finding all certain answers to queries w.r.t. materialized-view-settings, in all cases where the input instances are CQ weakly acyclic. In particular, we will see that the set $\mathcal{J}_{vv}^{\mathcal{M}\Sigma}$ is well defined, in that the chase tree in the view-verified data-exchange approach is always finite. We will also see that the set $\mathcal{J}_{vv}^{\mathcal{M}\Sigma}$ is "just tight enough," in the following sense: Recall (see Sect. 3.2) the definition of $certain_{\mathcal{M}\Sigma}(Q)$, i.e., of the set of certain answers of query Q w.r.t. materialized-view setting $\mathcal{M}\Sigma$. Then the expression in Eq. (4), which is the intersection of all the "certain-answer expressions" for Q and for the individual elements of the set $\mathcal{J}_{vv}^{\mathcal{M}\Sigma}$, is exactly the set $certain_{\mathcal{M}\Sigma}(Q)$.

Example 5. Recall the input instance $(\mathcal{M}\Sigma, Q)$ of Example 4, and the instance $J_{vv}^{\mathcal{M}\Sigma}$ obtained in that example. $J_{vv}^{\mathcal{M}\Sigma}$ is the (only) view-verified universal solution for $\mathcal{M}\Sigma$. The set of answers without nulls to the query Q on $J_{vv}^{\mathcal{M}\Sigma}$ is $\{(c, f), (g, f)\}$. The tuples (c, f) and (g, f) are certain answers of the query Q w.r.t. the materialized-view setting $\mathcal{M}\Sigma$, as computed for the instance $(\mathcal{M}\Sigma, Q)$ by the view-verified data-exchange approach.

5 Correctness, Validity, and Complexity of View-Verified Data Exchange

In this section, we show that the view-verified data-exchange approach is sound and complete for all CQ weakly acyclic input instances, and discuss its runtime and space complexity. We also show how the approach can be used to decide whether a CQ weakly acyclic materialized-view setting $\mathcal{M}\Sigma$ is valid.

5.1   View-Verified Data Exchange Is an Algorithm

We begin by obtaining a basic observation that builds on the results of [18] for chase with tgds and disjunctive egds (as they are defined in [18]). It is immediate from Proposition 3 that view-verified data exchange always terminates in finite time for CQ weakly acyclic inputs.

Proposition 3. *Given a CQ weakly acyclic materialized-view setting $\mathcal{M}\Sigma$, such that its canonical data-exchange solution $J_{de}^{\mathcal{M}\Sigma}$ exists. Assume that everything in $\mathcal{M}\Sigma$ is fixed except for the instance MV. Then we have that:*

(1) MV-enhanced chase of $J_{de}^{\mathcal{M}\Sigma}$ with $\Sigma \cup \Sigma^{(\mathcal{M}\Sigma)}$ is a finite tree, $\mathcal{T}$, such that:
   (a) $\mathcal{T}$ is of polynomial depth in the size of MV, and
   (b) The number of leaves in $\mathcal{T}$ is up to exponential in the size of MV; and
(2) For each nonempty leaf J of $\mathcal{T}$, we have that:
   (a) J is of polynomial size in the size of MV, and
   (b) Each grounded version of J is a Σ-valid base instance for $\mathcal{V}$ and MV.

(A *grounded version* of instance K results from replacing consistently all its nulls with distinct new constants.)

Proof (sketch). The proof of Proposition 3 relies heavily on the results of [18], particularly on its Theorem 3.9. Recall the "decomposition" of MV-induced generalized egds into egds as defined in Sect. 4.1. Intuitively, given a CQ weakly acyclic materialized-view setting $\mathcal{M}\Sigma$ and for each node K on each path from the root $J_{de}^{\mathcal{M}\Sigma}$ of the tree $\mathcal{T}$ for $\mathcal{M}\Sigma$, we can obtain K by chasing the root of $\mathcal{T}$ using only egds and weakly acyclic tgds.[4] The key observation here is that even though the set $\Sigma^{(\mathcal{M}\Sigma)}$ of dependencies is not fixed (in fact, its size is linear in the size of the instance MV in $\mathcal{M}\Sigma$), all the constants that contribute to the size of $\Sigma^{(\mathcal{M}\Sigma)}$ are already used in the root $J_{de}^{\mathcal{M}\Sigma}$ of the tree $\mathcal{T}$, by definition of $J_{de}^{\mathcal{M}\Sigma}$. In addition, the antecedent of each MV-induced generalized egd in $\Sigma^{(\mathcal{M}\Sigma)}$ is of constant size, by definition of the size of $\mathcal{M}\Sigma$. As a result, we can build on Theorem 3.9 and Proposition 5.6 of [18] to obtain items (1)(a) and (2)(a) of our Proposition 3.

Item (2)(b) of Proposition 3 is by definition of MV-enhanced chase, and (1)(b) is by construction of the tree $\mathcal{T}$. For the lower bound, please consider the following example, where for a CQ instance $\mathcal{M}\Sigma$ with $\Sigma = \emptyset$, the number of leaves in a chase tree is exponential in the size of $\mathcal{M}\Sigma$. (As usual and similarly to [29], we assume that the *size* of a given materialized-view setting $\mathcal{M}\Sigma$ is the size of its instance MV, with the remaining elements of $\mathcal{M}\Sigma$ being fixed.)

Example 6. Consider a schema $\mathbf{P}$ with two binary relations P and R, and with $\Sigma = \emptyset$. Let the set of views $\mathcal{V} = \{V, W\}$ be defined via two CQ queries, as follows:

---

[4] Besides the egds and tgds of Sect. 3.1, chase on each path in $\mathcal{T}$ may use MV-induced implication constraints. However, the only role of the latter constraints is to obtain the instance ϵ and thus to terminate the respective path in $\mathcal{T}$.

$V(X) \leftarrow P(X,Y), R(Y,Z).$
$W(Z) \leftarrow R(Y,Z).$

For each $n \geq 1$, consider a set $MV_{(n)}$ of answers for $\mathcal{V}$, with $n+2$ tuples, as follows. The relation $MV_{(n)}[V]$ has n tuples $V(1), V(2), \ldots, V(n)$, and $MV_{(n)}[W]$ has tuples $W(0)$ and $W(1)$.

For each $n \geq 1$, let the materialized-view setting $\mathcal{M}\Sigma^{(n)}$ be the tuple $(\mathbf{P}, \Sigma, \mathcal{V}, MV_{(n)})$, with all the components as described above. (As specified above, the set Σ is the empty set for each $n \geq 1$.)

The canonical universal solution $J_{de}^{\mathcal{M}\Sigma^{(n)}}$ for $MV_{(n)}$ has two tuples, $P(i, \perp_{(i,1)})$ and $R(\perp_{(i,1)}, \perp_{(i,2)})$, for $V(i)$ in $MV_{(n)}$, for each $i \in [1,n]$. It also has the tuples $R(\perp_{(n+1,1)}, 0)$ and $R(\perp_{(n+2,1)}, 1)$ for $MV_{(n)}[W]$.

The process of creating view-verified universal solutions for $\mathcal{M}\Sigma^{(n)}$ involves assigning either 0 or 1 independently to each of the nulls $\perp_{(i,2)}$, for all $i \in [1,n]$. It is easy to see that this process creates 2^n nonisomorphic instances, one for each assignment of zeroes and ones to each element of the vector $[\perp_{(1,2)}, \perp_{(2,2)}, \ldots, \perp_{(n,2)}]$. The expression 2^n is exponential in the size of the set $MV_{(n)}$ of view answers in $\mathcal{M}\Sigma^{(n)}$.

5.2   Soundness and Completeness

By Proposition 3 (2)(b), the view-verified data-exchange approach is a *complete* algorithm when applied to CQ weakly acyclic input instances $(\mathcal{M}\Sigma, Q)$. (That is, for each certain-answer tuple $\bar{t}$ for a problem input in this class, view-verified data exchange outputs $\bar{t}$.) We now make a key observation toward a proof that this algorithm is also *sound* for such instances. (Soundness means that for each tuple $\bar{t}$ that this approach outputs for an input $(\mathcal{M}\Sigma, Q)$ in this class, $\bar{t}$ is a certain-answer tuple for $(\mathcal{M}\Sigma, Q)$.)

Proposition 4. *Given a CQ weakly acyclic materialized-view setting* $\mathcal{M}\Sigma = (\mathbf{P}, \Sigma, \mathcal{V}, MV)$ *and a CQ query Q. Then, for each instance I such that* $\mathcal{V} \Rightarrow_{I,\Sigma} MV$, *there exists a homomorphism from some view-verified universal solution for* $\mathcal{M}\Sigma$ *to I.*

Proof (sketch). The intuition for the proof of Proposition 4 is as follows. For a given $\mathcal{M}\Sigma$, whenever an instance I exists such that $\mathcal{V} \Rightarrow_{I,\Sigma} MV$, a canonical data-exchange solution $J_{de}^{\mathcal{M}\Sigma}$ for $\mathcal{M}\Sigma$ must also exist. By definition of $J_{de}^{\mathcal{M}\Sigma}$, there must be a homomorphism from $J_{de}^{\mathcal{M}\Sigma}$ to the instance I. We then start applying MV-enhanced chase to $J_{de}^{\mathcal{M}\Sigma}$, to simulate some rooted path, $P_{(\mathcal{T})}$, in the chase tree $\mathcal{T}$ for $\mathcal{M}\Sigma$. (The tree is finite by Proposition 3.) In following the path $P_{(\mathcal{T})}$ via the chase, we make sure that there is a homomorphism from each node in the path to I, by always choosing an "appropriate" associated dependency $\tau^{(i)}$ for each MV-induced generalized egd τ that we are applying in the chase. By $\mathcal{V} \Rightarrow_{I,\Sigma} MV$, such a choice always exists, and the path $P_{(\mathcal{T})}$ terminates in finite time in a nonempty instance, J. By definition, J is a view-verified universal solution for $\mathcal{M}\Sigma$. By our simulation of the path $P_{(\mathcal{T})}$ "on the way to" I, there exists a homomorphism from J to I. □

5.3   Validity of Setting $\mathcal{M}\Sigma$

By the results of [18], when for a given $\mathcal{M}\Sigma$ no canonical data-exchange solution exists, then $\mathcal{M}\Sigma$ is not a valid setting. We refine this observation into a sufficient and necessary condition for validity of CQ weakly acyclic materialized-view settings $\mathcal{M}\Sigma$.

Proposition 5. *Given a CQ weakly acyclic materialized-view setting $\mathcal{M}\Sigma$, the setting $\mathcal{M}\Sigma$ is valid iff the set $\mathcal{J}_{vv}^{\mathcal{M}\Sigma}$ of view-verified universal solutions for $\mathcal{M}\Sigma$ is not empty.*

The only-if part of Proposition 5 is immediate from Proposition 4, and its if part is by Proposition 3 (2)(b).

5.4   Correctness of View-Verified Data Exchange

By Proposition 4, view-verified data exchange is sound. By Proposition 5, it outputs a set of certain-answer tuples iff its input is valid. We now conclude:

Theorem 4. *View-verified data exchange is a sound and complete algorithm for finding all certain answers to CQ queries w.r.t. CQ weakly acyclic materialized-view-settings.*

5.5   Complexity for CQ Weakly Acyclic Input Instances

By Theorem 4, view-verified data exchange is an algorithm for all CQ weakly acyclic input instances. We now obtain an exponential-time upper bound on the runtime complexity of the view-verified data-exchange approach, as follows.

Given a CQ weakly acyclic input instance $(\mathcal{M}\Sigma, Q)$, the runtime of the approach of Sect. 4.2 is exponential in the size of Q and of the set of answers MV in $\mathcal{M}\Sigma$, assuming that the rest of $\mathcal{M}\Sigma$ is fixed. This complexity setting extends naturally that of [29]: Zhang and Mendelzon in [29] assumed for their problem that the base schema and the view definitions are fixed, whereas the set of view answers MV and the queries posed on the base schema in presence of MV can vary. The authors of [29] did not consider dependencies on the base schema; we follow the standard data-exchange assumption, see, e.g., [18], that the given dependencies are fixed rather than being part of the problem input.

To obtain the above exponential-time upper bound for the problem of view-verified data exchange for CQ weakly acyclic input instances $(\mathcal{M}\Sigma, Q)$, we analyze the following flow for the view-verified data-exchange algorithm of Sect. 4.2. First, we spend exponential time in the arity k of Q to generate all the k-ary ground tuples $\bar{t}$ out of the set $consts(\mathcal{M}\Sigma)$. (Generating each such $\bar{t}$ gives rise to one iteration of the *main loop* of the algorithm.) For each such tuple $\bar{t}$, we then do the following:

- Construct the query $Q(\bar{t})$, as the result of applying to the query Q the homo-morphism[5] μ, such that (i) μ maps the head vector $\bar{X}$ of Q to $\bar{t}$, and (ii) μ is the identity mapping on each term that occurs in Q but not in its head vector $\bar{X}$;
- Enumerate all the (up to an exponential number of) view-verified universal solutions J for $\mathcal{MS}$ (recall that generating each such J takes polynomial time in the size of MV, see Proposition 3); and then
- For each such J that is not the empty instance, verify whether the query $Q(\bar{t})$ has a nonempty set of answers, which would be precisely $\{\bar{t}\}$, on the instance J. (For each $\bar{t}$ generated as above, we use a one-bit flag to track whether $\bar{t}$ is an answer to Q on all such instances J; each $\bar{t}$ that is an answer to Q on all the instances J is returned as an answer tuple by the view-verified data-exchange algorithm.) The runtime for this verification step is polynomial in the size of MV (because the size of J is polynomial in the size of MV, see Proposition 3) and is exponential in the number of subgoals of Q. (As the schema $\mathbf{P}$ in $\mathcal{MS}$ is fixed, each subgoal of the query Q has up to constant arity.)

By these observations, the following result holds:

Proposition 6. *The runtime complexity of the view-verified data-exchange algorithm with CQ weakly acyclic input instances is in EXPTIME.*

Observe that for each tuple $\bar{t}$ generated in the main loop of the algorithm, the respective iteration of the main loop runs in PSPACE. Indeed, recall from Proposition 3 that each instance J as above is of size polynomial in the size of the instance MV in $\mathcal{MS}$. Further, the size of each candidate valuation from $Q(\bar{t})$ to J is linear in the size of Q; thus, we satisfy the PSPACE requirement as long as we generate these candidate valuations one at a time ("on the fly" for each fixed J), in some clear algorithmic order.

Further, the entire view-verified data-exchange algorithm (i.e., finding *all* the certain-answer tuples for the given input CQ weakly acyclic pair $(\mathcal{MS}, Q)$) also runs in PSPACE, provided that we:

(a) Output each certain-answer tuple $\bar{t}$ "on the fly" (i.e., as soon as we know that it is a certain answer), and
(b) Use a counter (e.g., a binary-number representation of each k-ary ground candidate certain-answer tuple $\bar{t}$, as generated in the main loop of the algo-rithm) to keep track of the "latest" $\bar{t}$ that we have looked at and to generate from that "latest" $\bar{t}$ the next candidate certain-answer tuple $\bar{t}$ that we are to examine for the given input; the size of such a counter would be polynomial in the size of the problem input.

Thus, the following result holds:

Theorem 5. *The runtime complexity of the view-verified data-exchange algo-rithm with CQ weakly acyclic input instances is in PSPACE.*

---

[5] It is easy to verify that if a homomorphism μ specified by (i)-(ii) does not exist, then $\bar{t}$ cannot be a certain answer to Q w.r.t. $\mathcal{MS}$.

6 Related Work

The problem considered in this paper can be viewed as the problem of *inference control,* whose focus is on preventing unauthorized users from computing sensitive information. This problem has been studied extensively in database security, in both probabilistic (e.g., [24]) and deterministic versions. We consider the deterministic version of the problem. The reason is, when one can show that users can infer sensitive information in a conservative deterministic way, then the access-control rules are unsatisfactory and must be changed immediately. Work on deterministic inference control has been reported in [11,27], with solutions that can be made only sound but not complete for our information-leakage problem. To the best of our knowledge, none of the known procedures in inference control yields sound and complete algorithms for the information-leak disclosure problem.

Generally, the literature on privacy-preserving query answering and data publishing is represented by work on data anonymization and on differential privacy; please see [13] for a survey. Most of that work focuses on probabilistic inference of private information (see, e.g., [3]), while in this paper we focus on the possibilistic situation, where an adversary can deterministically derive sensitive information. (An interesting direction of future work is to see whether the approach of [3] can be combined with that of this current paper to address the probabilistic situation.) Further, our model of sensitive information goes beyond associations between individuals and their private sensitive attributes.

Policy analysis has been studied for various types of systems, including operating systems, role-based access control, trust management, and firewalls [5,6,20,23,25]. Typically, two types of properties are studied. The first type is static properties: Given the current security setting (e.g., non-management privileges of users), can certain actions or events (e.g., separation of duty) happen? Our analysis of information leakage falls into this category. What is different in our approach is that our policy model is much more elaborate, as we deal with policies defined by database query languages. The other type of properties in policy analysis is dynamic properties when a system evolves; that direction is not closely related to the topic of our paper.

The authors of [1] consider a problem whose statement is similar to ours. The results of [1] are different from ours, as we focus on developing constructive algorithms, rather than on doing complexity analysis as in [1]. Moreover, [1] use in their analysis a type of complexity metric that does not apply to data-access control. (The complexity metric that applies to our problem is that of [29].)

Our technical results build on the influential data-exchange framework of [18]. Our MV-induced dependencies of Sect. 4 resemble target-to-source dependencies Σ_{ts} introduced into (peer) data exchange in [19]. The difference is that Σ_{ts} are embedded dependencies defined at the schema level. In contrast, our MV-induced dependencies embody the given set of view answers MV.

7   Conclusion

Semantically seemingly benign access-control views may allow adversaries to infer sensitive information when considering specific answers to those views. We have seen that it may be easy enough for a human such as a database administrator (DBA) to see the leak of information. However, when given nontrivial security policies defined through views, the problem quickly becomes challenging. In particular, adversaries may obtain sensitive information through various inference methods. Known techniques for capturing specific inference methods often result in sound but incomplete solutions.

In this paper we studied the information-leakage problem that is fundamental in ensuring correct specification of data access-control policies in data-intensive systems. We formalized the problem and investigated two approaches to detecting information leakage. We showed that a naturally arising approach based on data exchange [7] is efficient and sound, but is incomplete. We then introduced a sound and complete view-verified data-exchange approach, which can be used as a foundation of detection tools for DBAs, and provided a study of its correctness and complexity in the context of CQ weakly acyclic input instances.

This paper reports on our first efforts in a formal study of information leakage in data. There are many important problems to investigate in our future work. In particular, in this paper we focused on the problem setting where a set MV of answers to access-control views (policies) is given. In the future work it is important to study the more general problem, which only considers integrity constraints and view definitions (policies) without a specific MV. It gives a much stronger security guarantee if we can show that information leakage cannot occur for all possible sets of view answers MV.

References

1. Abiteboul, S., Duschka, O.: Complexity of answering queries using materialized views. In: Proceedings of ACM SIGACT-SIGMOD-SIGART Symposium on Principles of Database Systems (PODS), pp. 254–263 (1998)
2. Abiteboul, S., Hull, R., Vianu, V.: Foundations of Databases. Addison-Wesley, Reading (1995)
3. Agrawal, R., Bayardo Jr., R.J., Faloutsos, C., Kiernan, J., Rantzau, R., Srikant, R.: Auditing compliance with a hippocratic database. In: Proceedings of Very Large Data Bases (VLDB) Conference, pp. 516–527 (2004)
4. Alborzi, F., Chirkova, R., Yu, T.: Exact detection of information leakage in database access control. In: Madria, S., Hara, T. (eds.) DaWaK 2015. LNCS, vol. 9263, pp. 403–415. Springer, Cham (2015). doi:10.1007/978-3-319-22729-0_31
5. Al-Shaer, E., Hamed, H., Boutaba, R., Hasan, M.: Conflict classification and analysis of distributed firewall policies. IEEE J. Sel. Areas Commun. **23**(10), 2069–2084 (2005)
6. Ammann, P., Sandhu, R.S.: Safety analysis for the extended schematic protection model. In: Proceedings of IEEE Symposium on Security and Privacy, pp. 87–97 (1991)

7. Barcelo, P.: Logical foundations of relational data exchange. SIGMOD Rec. **38**(1), 49–58 (2009)

8. Bertino, E., Ghinita, G., Kamra, A.: Access control for databases: concepts and systems. Found. Trends Databases **3**(1–2), 1–148 (2011)

9. Biskup, J., Bonatti, P.A.: Controlled query evaluation for known policies by combining lying and refusal. Ann. Math. Artif. Intell. **40**(1–2), 37–62 (2004)

10. Bond, R., See, K.Y.-K., Wong, C.K.M., Chan, Y.-K.H.: Understanding DB2 9 Security. IBM Press, Indianapolis (2006)

11. Brodsky, A., Farkas, C., Jajodia, S.: Secure databases: constraints, inference channels, and monitoring disclosures. IEEE Trans. Knowl. Data Eng. **12**(6), 900–919 (2000)

12. Chandra, A., Merlin, P.: Optimal implementation of conjunctive queries in relational data bases. In: Proceedings of the 9th Annual ACM Symposium on Theory of Computing (STOC), pp. 77–90 (1977)

13. Chen, B.-C., Kifer, D., LeFevre, K., Machanavajjhala, A.: Privacy-preserving data publishing. Found. Trends Databases **2**(1–2), 1–167 (2009)

14. Deutsch, A.: XML query reformulation over mixed and redundant storage. Ph.D. thesis, University of Pennsylvania (2002)

15. Deutsch, A., Nash, A., Remmel, J.: The chase revisited. In: Proceedings of ACM SIGACT-SIGMOD-SIGART Symposium on Principles of Database Systems (PODS), pp. 149–158 (2008)

16. Deutsch, A., Tannen, V.: Optimization properties for classes of conjunctive regular path queries. In: Ghelli, G., Grahne, G. (eds.) DBPL 2001. LNCS, vol. 2397, pp. 21–39. Springer, Heidelberg (2002). doi:10.1007/3-540-46093-4_2

17. Domingo-Ferrer, J.: Inference Control in Statistical Databases: From Theory to Practice. Springer, Heidelberg (2002)

18. Fagin, R., Kolaitis, P., Miller, R., Popa, L.: Data exchange: semantics and query answering. Theor. Comput. Sci. **336**(1), 89–124 (2005)

19. Fuxman, A., Kolaitis, P.G., Miller, R.J., Tan, W.-C.: Peer data exchange. ACM Trans. Database Syst. **31**(4), 1454–1498 (2006)

20. Harrison, M.A., Ruzzo, W.L., Ullman, J.D.: Protection in operating systems. Commun. ACM **19**, 461–471 (1976)

21. Kabra, G., Ramamurthy, R., Sudarshan, S.: Redundancy and information leakage in finite-grained access control. In: Proceedings of ACM SIGMOD International Conference on Management of Data, pp. 133–144 (2006)

22. Kaushik, R., Ramamurthy, R.: Efficient auditing for complex SQL queries. In: Proceedings of ACM SIGMOD International Conference on Management of Data, pp. 697–708 (2011)

23. Li, N., Winsborough, W.H., Mitchell, J.C.: Beyond proof-of-compliance: safety and availability analysis in trust management. In: Proceedings of IEEE Symposium on Security and Privacy, pp. 123–139 (2003)

24. Miklau, G., Suciu, D.: A formal analysis of information disclosure in data exchange. J. Comput. Syst. Sci. **73**(3), 507–534 (2007)

25. Motwani, R., Nabar, S., Thomas, D.: Auditing SQL queries. In: Proceedings of IEEE International Conference on Data Engineering (ICDE), pp. 287–296 (2008)

26. The Virtual Private Database in Oracle9iR2. An Oracle White Paper (2002)

27. Stoffel, K., Studer, T.: Provable data privacy. In: Andersen, K.V., Debenham, J., Wagner, R. (eds.) DEXA 2005. LNCS, vol. 3588, pp. 324–332. Springer, Heidelberg (2005). doi:10.1007/11546924_32

28. Zhang, X., Ozsoyoglu, M.: Implication and referential constraints: a new formal reasoning. IEEE Trans. Knowl. Data Eng. **9**(6), 894–910 (1997)
29. Zhang, Z., Mendelzon, A.O.: Authorization views and conditional query containment. In: Eiter, T., Libkin, L. (eds.) ICDT 2005. LNCS, vol. 3363, pp. 259–273. Springer, Heidelberg (2004). doi:10.1007/978-3-540-30570-5_18

Binary Shapelet Transform for Multiclass Time Series Classification

Aaron Bostrom[(✉)] and Anthony Bagnall

University of East Anglia, Norwich NR47TJ, UK
`a.bostrom@uea.ac.uk`

Abstract. Shapelets have recently been proposed as a new primitive for time series classification. Shapelets are subseries of series that best split the data into its classes. In the original research, shapelets were found recursively within a decision tree through enumeration of the search space. Subsequent research indicated that using shapelets as the basis for transforming datasets leads to more accurate classifiers. Both these approaches evaluate how well a shapelet splits all the classes. However, often a shapelet is most useful in 'distinguishing between members of the class of the series it was drawn from against all others. To assess this conjecture, we evaluate a one vs all encoding scheme. This technique simplifies the quality assessment calculations, speeds up the execution through facilitating more frequent early abandon and increases accuracy for multi-class problems. We also propose an alternative shapelet evaluation scheme which we demonstrate significantly speeds up the full search.

1 Introduction

Time series classification (TSC) is a subset of the general classification problem. The primary difference is that the ordering of attributes within each instance is important. For a set of n time series, $\mathbf{T} = \{T_1, T_2, ..., T_n\}$, each time series has m ordered real-valued observations, $T_i = \{t_{i1}, t_{i2}, ..., t_{im}\}$ and a class value c_i. The aim of TSC is to determine a function that relates the set of time series to the class values.

One recently proposed technique for TSC is to use shapelets [1]. Shapelets are subseries of the series $\mathbf{T}$ that best split the data into its classes. Shapelets can be used to detect discriminatory phase independent features that cannot be found with whole series measures such as dynamic time warping. Shapelet based classification involves measuring the similarity between a shapelet and each series, then using this similarity as a discriminatory feature for classification. The original shapelet-based classifier [1] embeds the shapelet discovery algorithm in a decision tree, and uses information gain to assess the quality of candidates. A shapelet is found at each node of the tree through an enumerative search. More recently, we proposed using shapelets as a transformation [2]. The shapelet transform involves a single-scan algorithm that finds the best k shapelets in a set of n time series. We use this algorithm to produce a transformed dataset,

© Springer-Verlag GmbH Germany 2017
A. Hameurlain et al. (Eds.): TLDKS XXXII, LNCS 10420, pp. 24–46, 2017.
DOI: 10.1007/978-3-662-55608-5_2

where each of the k features is the distance between the series and one shapelet. Hence, the value of the ith attribute of the jth record is the distance between the jth record and the ith shapelet. The primary advantages of this approach are that we can use the transformed data in conjunction with any classifier, and that we do not have to search sequentially for shapelets at each node. However, it still requires an enumerative search throughout the space of possible shapelets and the full search is $O(n^2m^4)$. Improvements for the full search technique were proposed in [1,3] and heuristics techniques to find approximations of the full search were described in [4–6].

Our focus is only on improving the exhaustive search. One of the problems of the shapelet search is the quality measures assess how well the shapelet splits all the classes. For multi-class problems measuring how well a shapelet splits all the classes may confound the fact that it actually represents a single class. Consider, for example, a shapelet in a data set of heartbeat measurements of patients with a range of medical conditions. It is more intuitive to imagine that a shapelet might represent a particular condition such as arrhythmia rather than discriminating between multiple conditions equally well. We redefine the transformation so that we find shapelets assessed on their ability to distinguish one class from all, rather than measures that separate all classes. This improves accuracy on multi-class problems and allows us to take greater advantage of the early abandon described in [1].

A further problem with the shapelet transform is that it may pick an excessive number of shapelets representing a single class. By definition, a good shapelet will appear in many series. The best way we have found to deal with this is to generate a large number of shapelets then cluster them [2]. However, there is still a risk that one class is generally easier to classify and hence has a disproportionate number of shapelets in the transform. The binary shapelet allows us to overcome this problem by balancing the number of shapelets we find for each class.

Finally, we describe an alternative way of enumerating the shapelet search that facilitates greater frequency of early abandon of the distance calculation.

2   Shapelet Based Classification

The shapelet transform algorithm described in [2] is summarised in Algorithm 1. Initially, for each time series, all candidates of length min to max are generated (i.e. extracted and normalised in the method $generateCandidates$). Then the distance between each shapelet and the other $n-1$ series are calculated to form the order list, D_S. Distance between a shapelet S of length l and a series T is given by

$$sDist(S,T) = \min_{w \in W_l} (dist(S,w)) \tag{1}$$

where W_l is the set of all l length subseries in T and $dist$ is the Euclidean distance between the equal length series S and w. The order list is used to determine the quality of the shapelet in the $assessCandidate$ method. Quality can be assessed by information gain [1] or alternative measures such as the F, moods median or rank order statistic [7]. Once all the shapelets for a series are evaluated they are sorted and the lowest quality overlapping shapelets are removed. The remaining candidates are then added to the shapelet set. By default, we set $k = 10n$ with the caveat that we do not accept shapelets that have zero information gain.

Algorithm 1. FullShapeletSelection($\mathbf{T}$, min, max, k)

Input: A list of time series $\mathbf{T}$, min and max length shapelet to search for and k, the maximum number of shapelets to find)

Output: A list of k shapelets

1:   $kShapelets \leftarrow \emptyset$
2:   **for all** T_i in $\mathbf{T}$ **do**
3:      $shapelets \leftarrow \emptyset$
4:      **for** $l \leftarrow min$ to max **do**
5:         $W_{i,l} \leftarrow generateCandidates(T_i, l)$
6:         **for all** subseries S in $W_{i,l}$ **do**
7:            $D_S \leftarrow findDistances(S, \mathbf{T})$
8:            $quality \leftarrow assessCandidate(S, D_S)$
9:            $shapelets.add(S, quality)$
10:     $sortByQuality(shapelets)$
11:     $removeSelfSimilar(shapelets)$
12:     $kShapelets \leftarrow merge(k, kShapelets, shapelets)$
13: **return** $kShapelets$

Once the best k shapelets have been found, the transform is performed with Algorithm 2. A more detailed description can be found in [8].

Extensions to the basic shapelet finding algorithm can be categorized into techniques to speed up the average case complexity of the exact technique and those that use heuristic search. The approximate techniques include reducing the dimensionality of the candidates and using a hash table to filter [4], searching the space of shapelet values (rather than taking the values from the train set series) [5] and randomly sampling the candidate shapelets [6]. Our focus is on improving the accuracy and speed of the full search. Two forms of early abandon described in [1] can improve the average case complexity. Firstly, the Euclidean distance calculations within the $sDist$ (Eq. 1) can be terminated early if they exceed the best found so far. Secondly, the shapelet evaluation can be abandoned early if $assessCandidate$ is updated as the $sDist$ are found and the best possible outcome for the candidate is worse than the current top candidates.

A speedup method involving trading memory for speed is proposed in [3]. For each pair of series T_i, T_j, cumulative sum, squared sum, and cross products of T_i and T_j are pre-calculated. With these statistics, the distance between sub-series can be calculated in constant time, making the shapelet-discovery algorithm $O(n^2m^3)$. However, pre-calculating of the cross products between all series prior to shapelet discovery requires $O(n^2m^2)$ memory, which is infeasible for most problems. Instead, [3] propose calculating these statistics prior to the start of the scan of each series, reducing the requirement to $O(nm^2)$ memory, but increasing the time overhead. Further refinements applicable to shapelets were described in [9], most relevant of which was a reordering of the sequence of calculations within the *dist* function to increase the likelihood of early abandon. The key observation is that because all series are normalized, the largest absolute values in the candidate series are more likely to contribute large values in the distance function. Hence, if the distances between positions with larger candidate values are evaluated first, then it is more likely the distance can be abandoned early. This can be easily implemented by creating an enumeration through the normalized candidate at the beginning, and adds very little overhead. We use this technique in all experiments.

Algorithm 2. *FullShapeletTransform*(Shapelets **S**,**T**)

1: $\mathbf{T}' \leftarrow \emptyset$
2: **for all** T in **T do**
3:     $T' \leftarrow <>$
4:     **for all** shapelets S in **S do**
5:         $dist \leftarrow sDist(S, T)$
6:         $T' \leftarrow append(T', dist)$
7:         $T' \leftarrow append(T', T.class)$
8:     $\mathbf{T}' \leftarrow \mathbf{T}' \cup T'$
9: **return T$'$**

---

3   Classification Technique

Once the transform is complete we can use any classifier on the problem. To reduce classifier induced variance we use a heterogenous ensemble of eight classifiers. The classifiers used are the WEKA [10] implementations of k Nearest Neighbour (where k is set through cross validation), Naive Bayes, C4.5 decision tree [11], Support Vector Machines [12] with linear and quadratic basis function kernels, Random Forest [13] (with 100 trees), Rotation Forest [14] (with 10 trees) and a Bayesian network. Each classifier is assigned a weight based on the cross validation training accuracy, and new data are classified with a weighted vote. The set of classifiers were chosen to balance simple and complex classifiers that use probabilistic, tree based and kernel based models. With the exception of k-NN, we do not optimise parameter settings for these classifiers via cross validation. More details are given in [8].

4   Alternate Shapelet Techniques

4.1   Fast Shapelets

The fast shapelet (FS) algorithm was proposed in 2013 [4]. The algorithm is a refinement of original decision tree shapelet selection algorithm. It employs a number of techniques to speed up the finding and pruning of shapelet candidates at each node of the tree [15]. The major changes made to the enumerative search is the introduction of symbolic aggregate approximation (SAX) [16] as a means for reducing the length of each series as well as smoothing and discretising the data. The other major advantage of using the SAX representation is that shapelet candidates can be pruned by using a collision table metric which highly correlates with Information Gain to reduce the amount of work performed in the quality measure stage. In the description in Algorithm 3 the decision tree has been omitted to improve clarity.

The first stage of the shapelet finding process is to create a list of SAX words [16]. The basic concept of SAX is a two stage process of dimension reduction and discretisation. SAX uses piece-wise aggregate approximation (PAA), to transform a time series into a number of smaller averaged sections before z normalization. This reduced series is discretised into a given alphabet size. Breakpoints are defined by equally likely areas of a standard normal distribution and each series forms a single string of characters. These strings are much smaller in length compared to the original series, so finding shapelets is faster. To increase the likelihood of word collisions a technique called random projects is employed. Given some SAX words random projection reduces their dimensionality by masking a number of their letters. The SAX words are randomly projected a number of times, the projected words are hashed and a frequency table for all the SAX words is built.

From this frequency table a new set of tables can be built which represent how common the SAX word is with respect to each class. A score for each SAX word can be calculated based on these grouping scores, and this value is used for assessing the distinguishing power of each SAX word. From this scoring process a list of the top k SAX shapelets can be created. These top k SAX shapelets are transformed back into their original series, where the shapelets information gain can be calculated. The best shapelet then forms the splitting rule in the decision tree.

Algorithm 3. FindBestShapelet(Set of time series **T**)

1: $bsfShapelet, shapelet$
2: $topK = 10$
3: **for** $length \leftarrow 5$ to m **do**
4:     $SAXList =$ FindSAXWords(**T**, $length$)
5:     RandomProjection($SAXList$)
6:     $ScoreList =$ ScoreAllSAX($SAXList$)
7:     $shapelet =$ FindBestSAX($ScoreList$, $SAXList$, $topK$)
8:     **if** $bsfShapelet < shapelet$ **then**
9:         $bsfShapelet = shapelet$
10: **return** $bsfShapelet$

4.2   Learn Shapelets

Learn shapelets (LS) was proposed by Grabocka *et al.* in 2014 [5]. Rather than uses subseries in the data as candidate shapelets, LS searches the space of all possible shapelets using a gradient descent approach on an initial set of shapelets found through clustering. An set of series is taken from the data and clustered using k-Means. The resulting centroids are refined with a two stage gradient descent model of shapelet refinement and logistic regression assessment. The learning is continued until either the model has converged or the number of iterations has exceeded a hard limit ($maxIter$). We show a high level overview of the algorithm presented in [5] in Algorithm 4.

Algorithm 4. LearnShapelets(Set of time series $\mathbf{T}$)

1: Parameters: $K, R, L_{min}, \eta, \lambda$
2: $\mathbf{S} \leftarrow$ InitKMeans($\mathbf{T}, K, R, L_{min}$)
3: $\mathbf{W} \leftarrow$ InitWeights($\mathbf{T}, K, R$)
4: **for** $i \leftarrow maxIter$ **do**
5:    $\mathbf{M} \leftarrow$ updateModel($\mathbf{T}, \mathbf{S}, \alpha, L_{min}, R$)
6:    $\mathbf{L} \leftarrow$ updateLoss($\mathbf{T}, \mathbf{M}, \mathbf{W}$)
7:    $\mathbf{W}, \mathbf{S} \leftarrow$ updateWandS($\mathbf{T}, \mathbf{M}, \mathbf{W}, \mathbf{S}, \eta, R, L_{min}, L, \lambda_W, \alpha$)
8:    **if** diverged() **then**
9:        $i = 0$
10:       $\eta = \eta/3$

---

5   Shapelet Transform Refinements

5.1   Binary Shapelets

The standard shapelet assessment method measures how well the shapelet splits up all the classes. There are three potential problems with this approach when classifying multi-class problems. The problems apply to all possible quality measures, but we use information gain to demonstrate the point. Firstly, useful information about a single class may be lost. For example, suppose we have a four class problem and a shapelet produces the order line presented in Fig. 1, where each colour represents a different class.

The first shapelet groups all of class 1 very well, but cannot distinguish between classes 2, 3 and 4 and hence has a lower information gain than the split produced by the second shapelet in Fig. 1 which separates class 1 and 2 from class 3 and 4. The more classes there are, the more likely it is that the quantification of the ability of a shapelet to separate out a single class will be overwhelmed by the mix of other class values. We can mitigate against this potential problem by defining a binary shapelet as one that is assessed by how well it splits the class of the series it originated from all the other classes. The second problem with

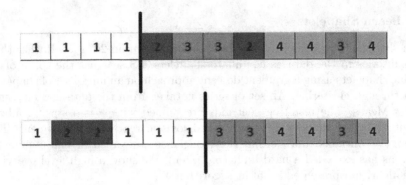

Fig. 1. An example order line split for two shapelets. The top shapelet discriminates between class 1 and the rest perfectly, yet has lower information gain than the orderline shown below it.

searching all shapelets with multi-class assessment arises if one class is much easier to classify than the others. In this case it is likely that more shapelets will be found for the easy class than for the other classes. Although our principle is to find a large number of shapelets (ten times the number of training cases) and let the classifier deal with redundant features, there is still a risk that a large number of similar shapelets for one class will crowd out useful shapelets for another class. If we use binary shapelets we can simply allocate a maximum number of shapelets to each class. We adopt the simple approach of allocating a maximum of k/c shapelets to each class, where c is the number of classes. The final problem is that the shapelet early abandon described in [1] is not useful for multi-class problems. Given a partial orderline and a split point, the early abandon works by upper bounding the information gain by assigning the unassigned series to the side of the split that would give the maximum gain. However, the only way to do this with multi-class problems is to try all permutations. The time this takes quickly rises to offset the possible benefits from the early abandon. If we restrict our attention to just binary shapelets then we can take maximum advantage of the early abandon. The binary shapelet selection is described by Algorithm 5.

5.2 Changing the Shapelet Evaluation Order

Shapelets are phase independent. However, for many problems the localised features are at most only weakly independent in phase, i.e. the best matches will appear close to the location of the candidate. Finding a good match early in $sDist$ increases the likelihood of an early abandon for each $dist$ calculation. Hence, we redefine the order of iteration of the $dist$ calculations within $sDist$ so that we start with the index the shapelet was found at and move consecutively left and right from that point. Figure 2 demonstrates the potential benefit of this approach. The scan from the beginning is unable to early abandon on any of the subseries before the best match. The scan originating at the candidates location finds the best match faster an hence can early abandon on all the distance calcu-

---

Algorithm 5. BinaryShapeletSelection(**T**, min, max, k)

Input: A list of time series **T**, min and max length shapelet to search for and k,the maximum number of shapelets to find)
Output: A list of k Shapelets
1: $numClasses \leftarrow getClassDistribution(T)$
2: $kShapeletsMap \leftarrow \emptyset$
3: $prop \leftarrow k/numClasses$
4: **for all** T_i in **T do**
5:     $shapelets \leftarrow \emptyset$
6:     **for** $l \leftarrow min$ to max **do**
7:         $W_{i,l} \leftarrow generateCandidates(T_i, l)$
8:         **for all** subseries S in $W_{i,l}$ **do**
9:             $D_S \leftarrow findDistances(S, \mathbf{T})$
10:             $quality \leftarrow assessCandidate(S, D_S)$
11:             $shapelets.add(S, quality)$
12:     $sortByQuality(shapelets)$
13:     $removeSelfSimilar(shapelets)$
14:     $kShapelets \leftarrow kShapeletsMap.get(T.class)$
15:     $kShapelets \leftarrow merge(prop, kShapelets, shapelets)$
16:     $kShapeletsMap.add(kShapelets, T.class)$
17: **return** $kShapeletsMap.asList()$

---

lations at the beginning of the series. Hence, if the location of the best shapelet is weakly phase dependent, we would expect to observe an improvement in the time complexity. The revised function $sDist$, which is a subroutine of $findDistances$ (line 9 in Algorithm 5), is described in Algorithm 6.

6   Results

We demonstrate the utility of our approach through experiments using 74 benchmark multi-class datasets from the UCR Time Series Classification archive [17]. In common with the vast majority of research in this field, we present results on the standard train/test split. The min and max size of the shapelet are set to 3 and m (series length). As a sanity check, we have also evaluated the binary shapelets on two class problems to demonstrate there is negligible difference. On 25 two class problems, the full transform was better on 6, the binary transform better on 19 and they were tied on 1. All the results and the code to generate them are available from [18,19].

6.1   Accuracy Improvement on Multi-class Problems

Table 1 gives the results for the full shapelet transform and the binary shapelet transform on problems with 2-50 classes. Overall, the binary shapelet transform is better on 48 data sets, the full transform better on 21 and on 4 they are equal. On multi class problems, the binary shapelet transform is better on 29 problems

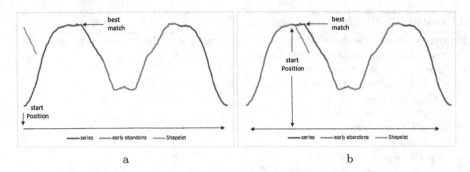

a   b

Fig. 2. An example of Euclidean distance early abandon where the *sDist* scan starts from the beginning (a) and from the place of origin of the candidate shapelet (b). For the scan from the beginning, there are no early abandons until the scan has passed the best match. Because the best match is close to the location of the candidate shapelet, starting from the shapelets original location allows for a greater number of early abandons.

Algorithm 6. $sDist$(shapelet S, series T_i)

1: $subSeq \leftarrow getSubSeq(T_i, S.startPos, S.length)$
2: $bestDist \leftarrow euclideanDistance(subSeq, S)$
3: $i \leftarrow 1$
4: **while** $leftExists \parallel rightExists$ **do**
5:    $leftExists \leftarrow S.startPos - i \geq 0$
6:    $rightExists \leftarrow S.startPos + i < T_i.length$
7:    **if** $rightExists$ **then**
8:       $subSeq \leftarrow getSubSeq(T_i, S.startPos + i, S.length)$
9:       $currentDist \leftarrow earlyAbandonDistance(subSeq, S, bestDist)$
10:       **if** $currentDist > bestDist$ **then**
11:          $bestDist \leftarrow currentDist$
12:    **if** $leftExists$ **then**
13:       $subSeq \leftarrow getSubSeq(T_i, S.startPos - i, S.length)$
14:       $currentDist \leftarrow earlyAbandonDistance(subSeq, S, bestDist)$
15:       **if** $currentDist > bestDist$ **then**
16:          $bestDist \leftarrow currentDist$
17:    $i \leftarrow i + 1$
18: **return** $bestDist$

---

compared with the full shapelet transform being better on 15. The difference between the full and the binary shapelet transform is significant at the 5% level using a paired T test.

Figure 3 shows the plot of the difference in accuracy of the full and binary shapelet transform plotted against the number of classes. There is a clear trend of increasing accuracy for the binary transform as the number of classes continues. This is confirmed in Table 2, which presents the same data grouped into bins of ranges of number of classes.

Table 1. Full shapelet transform vs. Binary shapelet transform.

Dataset	#Classes	FST	binaryST
Adiac	37	0.565	**0.783**
ArrowHead	3	**0.771**	0.737
Beef	5	0.833	**0.9**
BeetleFly	2	**0.75**	0.6
BirdChicken	2	0.75	**0.8**
Car	4	0.733	**0.917**
CBF	3	**0.997**	0.974
ChlorineConcentration	3	**0.7**	0.7
CinCECGtorso	4	0.846	**0.954**
Coffee	2	**1**	0.964
Computers	2	0.7	**0.736**
CricketX	12	**0.782**	0.772
CricketY	12	0.764	**0.779**
CricketZ	12	0.772	**0.787**
DiatomSizeReduction	4	0.876	**0.925**
DistalPhalanxOutlineAgeGroup	3	0.741	**0.77**
DistalPhalanxOutlineCorrect	2	0.736	**0.775**
DistalPhalanxTW	6	0.633	**0.662**
Earthquakes	2	0.734	**0.741**
ECGFiveDays	2	**0.999**	0.984
FaceAll	14	0.737	**0.779**
FaceFour	4	**0.943**	0.852
FacesUCR	14	**0.913**	0.906
fiftywords	50	**0.719**	0.705
fish	7	0.977	**0.989**
FordA	2	0.927	**0.971**
FordB	2	0.789	**0.807**
GunPoint	2	0.98	**1**
Haptics	5	0.477	**0.523**
Herring	2	**0.672**	0.672
InlineSkate	7	**0.385**	0.373
ItalyPowerDemand	2	**0.952**	0.948
LargeKitchenAppliances	3	**0.883**	0.859
Lightning2	2	0.656	**0.738**
Lightning7	7	**0.74**	0.726
MALLAT	8	0.94	**0.964**

(Continued)

Table 1. (Continued)

Dataset	#Classes	FST	binaryST
MedicalImages	10	0.604	**0.67**
MiddlePhalanxOutlineAgeGroup	3	0.63	**0.643**
MiddlePhalanxOutlineCorrect	2	0.725	**0.794**
MiddlePhalanxTW	6	**0.539**	0.519
MoteStrain	2	0.891	**0.897**
NonInvasiveFatalECGThorax1	42	0.9	**0.95**
NonInvasiveFatalECGThorax2	42	0.903	**0.951**
OliveOil	4	**0.9**	0.9
OSULeaf	6	0.715	**0.967**
PhalangesOutlinesCorrect	2	0.748	**0.763**
Plane	7	**1**	1
ProximalPhalanxOutlineAgeGroup	3	**0.854**	0.844
ProximalPhalanxOutlineCorrect	2	**0.9**	0.883
ProximalPhalanxTW	6	0.771	**0.805**
RefrigerationDevices	3	0.557	**0.581**
ScreenType	3	**0.533**	0.52
ShapeletSim	2	0.919	**0.956**
SmallKitchenAppliances	3	0.773	**0.792**
SonyAIBORobotSurface1	2	**0.933**	0.864
SonyAIBORobotSurface2	2	0.885	**0.934**
StarLightCurves	3	0.976	**0.979**
SwedishLeaf	15	0.907	**0.928**
Symbols	6	**0.886**	0.882
SyntheticControl	6	**0.983**	0.983
ToeSegmentation1	2	0.956	**0.965**
ToeSegmentation2	2	0.854	**0.908**
Trace	4	0.98	**1**
TwoLeadECG	2	0.996	**0.997**
TwoPatterns	4	0.941	**0.955**
UWaveGestureLibraryX	8	0.784	**0.803**
UWaveGestureLibraryY	8	0.697	**0.73**
UWaveGestureLibraryZ	8	0.727	**0.748**
wafer	2	0.998	**1**
WordSynonyms	25	**0.597**	0.571
Worms	5	0.701	**0.74**
WormsTwoClass	2	0.766	**0.831**
Yoga	2	0.805	**0.818**
Total wins		21	48

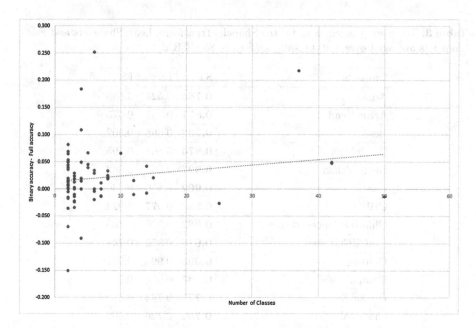

Fig. 3. Number of classes plotted against the difference in error between the full shapelets and the binary shapelets. A positive number indicates the binary shapelets are better. The dotted line is the least squares regression line.

6.2 Accuracy Comparison to Other Shapelet Methods

To establish the efficacy of the binary shapelet approach we wanted to thoroughly compare it to other shapelet approaches within the literature, Fast Shapelets (FS) [4] and Learn Shapelets (LS) [5]. The 85 datasets from the updated repository [18] were stratified and resampled 100 times, to produce 8500 problems. All folds and results are reproducible within our common Java WEKA [10] framework [19]. Where possible we tried to match the experimental procedure for parameter searching outlined in the original work. We performed very extensive tests and analysis on the algorithms implemented to make sure they were

Table 2. Number of data sets the binary shapelet beats the full shapelet split by number of classes.

Number of classes	Full better	Binary better
2	6	19
3 to 5	7	13
6 to 9	4	8
10 and above	4	8
All	21	48

Table 3. The average accuracies for the ShapeletTransform, LearnShapelets and Fast-Shapelets averaged over a 100 resamples for the 85 UCR datasets

Datasets	ST	LS	FS
Adiac	**0.768**	0.527	0.555
ArrowHead	**0.851**	0.841	0.675
Beef	**0.736**	0.698	0.502
BeetleFly	**0.874**	0.861	0.795
BirdChicken	**0.927**	0.863	0.862
Car	**0.902**	0.856	0.736
CBF	**0.986**	0.977	0.924
ChlorineConcentration	**0.682**	0.586	0.566
CinCECGtorso	**0.918**	0.855	0.741
Coffee	**0.995**	0.995	0.917
Computers	**0.785**	0.654	0.5
CricketX	**0.777**	0.744	0.479
CricketY	**0.762**	0.726	0.509
CricketZ	**0.798**	0.754	0.466
DiatomSizeReduction	0.911	**0.927**	0.873
DistalPhalanxOutlineCorrect	**0.829**	0.822	0.78
DistalPhalanxOutlineAgeGroup	**0.819**	0.81	0.745
DistalPhalanxTW	**0.69**	0.659	0.623
Earthquakes	0.737	0.742	**0.747**
ECG200	0.84	**0.871**	0.806
ECG5000	**0.943**	0.94	0.922
ECGFiveDays	0.955	0.985	**0.986**
ElectricDevices	**0.895**	0.709	0.262
FaceAll	**0.968**	0.926	0.772
FaceFour	0.794	**0.957**	0.869
FacesUCR	0.909	**0.939**	0.701
FiftyWords	**0.713**	0.694	0.512
Fish	**0.974**	0.94	0.742
FordA	**0.965**	0.895	0.785
FordB	**0.915**	0.89	0.783
GunPoint	**0.999**	0.983	0.93
Ham	0.808	**0.832**	0.677
HandOutlines	**0.924**	0.837	0.841
Haptics	**0.512**	0.478	0.356

(Continued)

Table 3. *(Continued)*

Datasets	ST	LS	FS
Herring	**0.653**	0.628	0.558
InlineSkate	**0.393**	0.299	0.257
InsectWingbeatSound	**0.617**	0.55	0.488
ItalyPowerDemand	**0.953**	0.952	0.909
LargeKitchenAppliances	**0.933**	0.765	0.419
Lightning2	0.659	**0.759**	0.48
Lightning7	0.724	**0.765**	0.101
Mallat	**0.972**	0.951	0.893
Meat	**0.966**	0.814	0.924
MedicalImages	0.691	**0.704**	0.609
MiddlePhalanxOutlineCorrect	0.815	**0.822**	0.716
MiddlePhalanxOutlineAgeGroup	**0.694**	0.679	0.613
MiddlePhalanxTW	**0.579**	0.54	0.519
MoteStrain	**0.882**	0.876	0.793
NonInvasiveFatalECGThorax1	**0.947**	0.6	0.71
NonInvasiveFatalECGThorax2	**0.954**	0.739	0.758
OliveOil	**0.881**	0.172	0.765
OSULeaf	**0.934**	0.771	0.679
PhalangesOutlinesCorrect	**0.794**	0.783	0.73
Phoneme	**0.329**	0.152	0.173
Plane	**1**	0.995	0.97
ProximalPhalanxOutlineCorrect	**0.881**	0.793	0.797
ProximalPhalanxOutlineAgeGroup	**0.841**	0.832	0.797
ProximalPhalanxTW	**0.803**	0.794	0.716
RefrigerationDevices	**0.761**	0.642	0.574
ScreenType	**0.676**	0.445	0.365
ShapeletSim	0.934	0.933	**1**
ShapesAll	**0.854**	0.76	0.598
SmallKitchenAppliances	**0.802**	0.663	0.333
SonyAIBORobotSurface1	0.888	0.906	**0.918**
SonyAIBORobotSurface2	**0.924**	0.9	0.849
StarlightCurves	**0.977**	0.888	0.908
Strawberry	**0.968**	0.925	0.917
SwedishLeaf	**0.939**	0.899	0.758

(Continued)

Table 3. *(Continued)*

Datasets	ST	LS	FS
Symbols	0.862	**0.919**	0.908
SyntheticControl	0.987	**0.995**	0.92
ToeSegmentation1	**0.954**	0.934	0.904
ToeSegmentation2	**0.947**	0.943	0.873
Trace	1	0.996	0.998
TwoLeadECG	0.984	**0.994**	0.92
TwoPatterns	0.952	**0.994**	0.696
UWaveGestureLibraryX	**0.806**	0.804	0.694
UWaveGestureLibraryY	**0.737**	0.718	0.591
UWaveGestureLibraryZ	**0.747**	0.737	0.638
UWaveGestureLibraryAll	**0.942**	0.68	0.766
Wafer	1	0.996	0.981
Wine	**0.926**	0.524	0.794
WordSynonyms	**0.582**	0.581	0.461
Worms	**0.719**	0.642	0.622
WormsTwoClass	**0.779**	0.736	0.706
Yoga	0.823	**0.833**	0.721
Total wins	71	14	4

identical to the available source code, and provide statistically similar results as published. This often required working with the original authors to help replicate and fix errors.

We present the results of the mean accuracy for the binary shapelet transform, FS and LS in Table 3. We found that binary shapelets was better than FS and LS on 67 problems, and on a pair wise t-test was significantly better at the 5% level. We show the critical difference of the three approaches in Fig. 4.

6.3   Accuracy Comparison to Standard Approaches

Using the same methodology for comparing shapelet methodsm we compared the shapelet approach to more standard approaches. These were 1-nearest neighbour using Euclidean distance, 1 nearest neighbour with dynamic time warping and lastly rotation forest. We present the results in Table 4 showing that on the 85 datasets, the binary shapelet wins on 55 problems and is significantly better than other the standard approaches. We show the comparison of these classifiers in the critical difference diagram in Fig. 5.

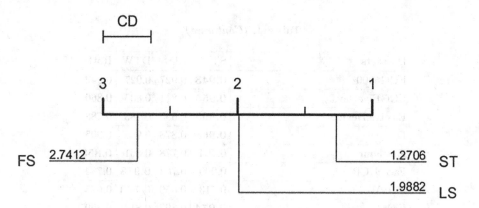

Fig. 4. The critical difference diagram of the Shapelet Transform, Fast Shapelets and Learn Shapalets, the data is presented in 3

Table 4. The average accuracies for the BinaryShapeletTransform, 1NN with Euclidean distance, 1NN with DTW setting window size through cross-validation and Rotation Forest, averaged over a 100 resamples for the 85 UCR datasets

Datasets	ST	ED	DTW	RotF
Adiac	**0.768**	0.617	0.615	0.754
ArrowHead	**0.851**	0.841	0.829	0.789
Beef	0.736	0.533	0.532	**0.819**
BeetleFly	**0.875**	0.686	0.785	0.791
BirdChicken	**0.927**	0.693	0.823	0.747
Car	**0.902**	0.724	0.714	0.788
CBF	**0.986**	0.870	0.974	0.898
ChlorineConcentration	0.682	0.652	0.651	**0.846**
CinCECGtorso	0.918	0.891	**0.928**	0.712
Coffee	**0.995**	0.981	0.981	0.995
Computers	**0.785**	0.575	0.688	0.666
CricketX	**0.777**	0.579	0.774	0.620
CricketY	**0.762**	0.545	0.749	0.599
CricketZ	**0.798**	0.589	0.779	0.626
DiatomSizeReduction	0.911	0.943	**0.944**	0.881
DistalPhalanxOutlineCorrect	**0.829**	0.744	0.756	0.812
DistalPhalanxOutlineAgeGroup	**0.819**	0.731	0.733	0.807
DistalPhalanxTW	0.690	0.628	0.621	**0.692**
Earthquakes	0.737	0.682	0.696	**0.759**
ECG200	0.840	**0.879**	0.872	0.851

(Continued)

Table 4. *(Continued)*

Datasets	ST	ED	DTW	RotF
ECG5000	**0.943**	0.927	0.927	0.942
ECGFiveDays	**0.955**	0.811	0.835	0.860
ElectricDevices	**0.895**	0.695	0.783	0.788
FaceAll	**0.968**	0.878	0.958	0.905
FaceFour	0.794	0.778	0.851	**0.853**
FacesUCR	0.909	0.764	**0.918**	0.784
FiftyWords	0.713	0.659	**0.770**	0.675
Fish	**0.974**	0.802	0.814	0.859
FordA	**0.965**	0.682	0.684	0.837
FordB	**0.915**	0.648	0.661	0.808
GunPoint	**0.999**	0.891	0.947	0.924
Ham	0.808	0.758	0.749	**0.822**
HandOutlines	**0.924**	0.853	0.857	0.912
Haptics	**0.512**	0.386	0.409	0.469
Herring	**0.653**	0.496	0.545	0.608
InlineSkate	0.393	0.326	**0.400**	0.340
InsectWingbeatSound	0.617	0.553	0.555	**0.633**
ItalyPowerDemand	0.953	0.954	0.951	**0.967**
LargeKitchenAppliances	**0.933**	0.524	0.788	0.622
Lightning2	0.659	0.716	**0.831**	0.760
Lightning7	0.724	0.618	**0.744**	0.701
Mallat	**0.972**	0.933	0.943	0.946
Meat	0.966	0.981	0.980	**0.994**
MedicalImages	0.691	0.701	0.748	**0.756**
MiddlePhalanxOutlineCorrect	0.815	0.776	0.775	**0.820**
MiddlePhalanxOutlineAgeGroup	**0.694**	0.583	0.570	0.669
MiddlePhalanxTW	**0.579**	0.493	0.496	0.568
MoteStrain	**0.882**	0.866	0.862	0.859
NonInvasiveFatalECGThorax1	**0.947**	0.822	0.820	0.899
NonInvasiveFatalECGThorax2	**0.954**	0.889	0.884	0.928
OliveOil	0.881	0.877	0.876	**0.889**
OSULeaf	**0.934**	0.573	0.634	0.587
PhalangesOutlinesCorrect	0.794	0.768	0.766	**0.833**
Phoneme	**0.329**	0.104	0.230	0.127

(Continued)

Table 4. *(Continued)*

Datasets	ST	ED	DTW	RotF
Plane	**1.000**	0.967	0.994	0.986
ProximalPhalanxOutlineCorrect	**0.881**	0.818	0.816	0.875
ProximalPhalanxOutlineAgeGroup	0.841	0.770	0.765	**0.847**
ProximalPhalanxTW	0.803	0.713	0.732	**0.808**
RefrigerationDevices	**0.761**	0.426	0.573	0.570
ScreenType	**0.676**	0.432	0.465	0.466
ShapeletSim	**0.934**	0.505	0.652	0.488
ShapesAll	**0.854**	0.754	0.804	0.760
SmallKitchenAppliances	**0.802**	0.370	0.674	0.714
SonyAIBORobotSurface1	**0.888**	0.785	0.804	0.814
SonyAIBORobotSurface2	**0.924**	0.855	0.855	0.846
StarlightCurves	**0.977**	0.853	0.908	0.970
Strawberry	0.968	0.956	0.955	**0.974**
SwedishLeaf	**0.939**	0.772	0.842	0.884
Symbols	0.862	0.876	**0.920**	0.842
SyntheticControl	0.987	0.903	**0.989**	0.967
ToeSegmentation1	**0.954**	0.612	0.722	0.578
ToeSegmentation2	**0.947**	0.781	0.851	0.646
Trace	**1.000**	0.778	0.993	0.932
TwoLeadECG	**0.984**	0.735	0.881	0.928
TwoPatterns	0.952	0.906	**0.999**	0.928
UWaveGestureLibraryX	**0.806**	0.740	0.777	0.779
UWaveGestureLibraryY	**0.737**	0.665	0.695	0.717
UWaveGestureLibraryZ	**0.747**	0.661	0.687	0.728
UWaveGestureLibraryAll	0.942	0.943	**0.960**	0.946
Wafer	**1.000**	0.995	0.995	0.995
Wine	**0.926**	0.893	0.891	0.919
WordSynonyms	0.582	0.615	**0.730**	0.586
Worms	**0.719**	0.491	0.569	0.605
WormsTwoClass	**0.779**	0.624	0.661	0.657
Yoga	0.823	0.840	**0.858**	0.854
Total wins	55	1	13	16

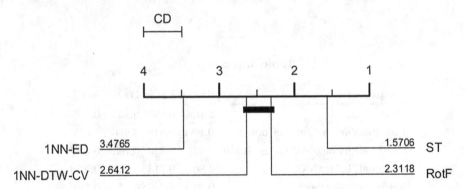

Fig. 5. The critical difference diagram showing the comparison of the new Shapelet Transform compared with a number of standard benchmark classifiers; 1 Nearest Niegh-bour with Euclidean Distance, 1 Nearest Neighbour with Dynamic Time Warping setting the windows size through cross-validation and Rotation Forest, results are shown in Table 4

6.4   Average Case Time Complexity Improvements

One of the benefits of using the binary transform is that it is easier to use the shapelet early abandon described in [3]. Early abandon is less useful when finding the best k shapelets than it is for finding the single best, but when it can be employed it can give real benefit. Figure 6 shows that on certain datasets, using the binary shapelet discovery means millions of fewer $sDist$ evaluations.

We assess the improvement from using Algorithm 6 by counting the number of point wise distance calculations required from using the standard approach, the alternative enumeration, and the state of art the enumeration in [2].

For the datasets used in our accuracy experiments, changing the order of enumeration reduces the number of calculations in the distance function by 76% on average. The improvement ranges from negligible (e.g. Lightning7 requires 99.3% of the calculations) to substantial (e.g. Adiac operations count is 63% of the standard approach). This highlights that the best shapelets may or may not be phase independent, but nothing is lost from changing the evaluation order and often substantial improvements are achieved. This is further highlighted in Fig. 7 where the worst dataset Synthetic control does not benefit from the alternate enumeration. For the average and the best case we see a reduction of approx. 15% fewer operations required. On the Olive Oil dataset we see a 99% reduction in the number of distance calculations required.

Full results, all the code and the datasets used can be downloaded from [18,19]. The set of experiments, and results are constantly being expanded and evaluated against the current state of the art.

We define the opCounts as the number of operations performed in the Euclidean distance function when comparing two series. This enables us to estimate the improvements of new techniques. We show in Table 5 the op counts in millions for 40 datasets. The balancing in some rare cases can increase the number of

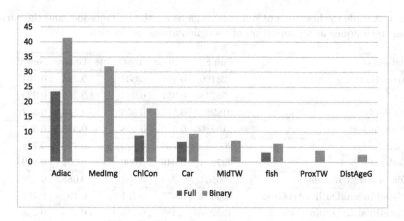

Fig. 6. Number of $sDist$ measurements that were not required because of early abandon (in millions) for both full and binary shapelet discovery on seven datasets.

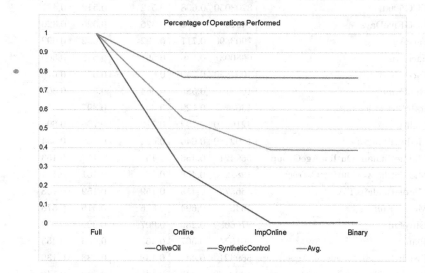

Fig. 7. Percentage of operations performed of Full search, compared with previous sDist optimisations, and our two new speed up techniques.

operations performed. Typically these datasets have a large number of a class swamping the shapelet set. The balancer has individual lists for each class, and thus the shapelet being compared to for early abandon entropy pruning is different for each class, and in some cases can mean more work is done. In these experiments we wanted to show the progression of op count reduction as we combined techniques. When comparing to the previous best early abandon technique, the combination of the new methods proposed has seen a reduction of on average 35% of the required operations, to find and evaluate the same shapelets.

Table 5. A table to show the opCounts in millions for the Full method, and the current and new techniques as a proportion of the calculations performed

Datasets	Full ST	Online	ImpOnline	Binary	New ST
ArrowHead	423196	0.496	0.307	0.298	0.301
Beef	3567380	0.357	0.234	0.266	0.285
BeetleFly	2192860	0.576	0.555	0.555	0.557
BirdChicken	2192860	0.475	0.398	0.398	0.403
CBF	20033	0.768	0.726	0.726	0.726
Coffee	427243	0.449	0.090	0.090	0.089
DiatomSizeReduction	286551	0.364	0.088	0.088	0.088
DistalPhalanxOutlineAgeGroup	569421	0.475	0.191	0.191	0.187
DistalPhalanxOutlineCorrect	1282270	0.489	0.218	0.218	0.218
DistalPhalanxTW	569421	0.474	0.192	0.192	0.178
ECG200	72759	0.568	0.448	0.448	0.448
ECG5000	8203050	0.608	0.512	0.512	0.487
ECGFiveDays	14826	0.580	0.326	0.326	0.326
FaceAll	7903390	0.717	0.692	0.692	0.692
FaceFour	698003	0.686	0.561	0.566	0.566
FacesUCR	1004840	0.740	0.676	0.676	0.670
GunPoint	105975	0.585	0.364	0.364	0.364
ItalyPowerDemand	136	0.529	0.397	0.397	0.397
Lightning7	4219010	0.732	0.712	0.706	0.696
MedicalImages	1202180	0.640	0.622	0.622	0.534
MiddlePhalanxOutlineAgeGroup	569421	0.456	0.156	0.156	0.153
MiddlePhalanxOutlineCorrect	1282270	0.458	0.161	0.161	0.161
MiddlePhalanxTW	566573	0.453	0.159	0.159	0.158
MoteStrain	1645	0.694	0.576	0.576	0.578
OliveOil	7706080	0.280	0.007	0.009	0.009
Plane	401574	0.567	0.455	0.455	0.455
ProximalPhalanxOutlineAgeGroup	569421	0.441	0.138	0.138	0.136
ProximalPhalanxOutlineCorrect	1282270	0.447	0.140	0.140	0.140
ProximalPhalanxTW	569421	0.443	0.138	0.138	0.137
ShapeletSim	1994760	0.690	0.673	0.673	0.679
SonyAIBORobotSurface1	799	0.579	0.397	0.397	0.397
SonyAIBORobotSurface2	1101	0.654	0.561	0.561	0.561
SwedishLeaf	5745210	0.550	0.393	0.393	0.389
Symbols	1266960	0.632	0.601	0.601	0.601
SyntheticControl	102522	0.771	0.768	0.768	0.768
ToeSegmentation1	776099	0.620	0.611	0.611	0.607
ToeSegmentation2	1469900	0.555	0.556	0.556	0.554
Trace	4785000	0.691	0.634	0.634	0.634
TwoLeadECG	1991	0.497	0.203	0.203	0.203
Wine	810711	0.383	0.023	0.023	0.023

7 Conclusion

Shapelets are useful for classifying time series where the between class variability can be detected by relatively short, phase independent subseries. They offer an alternative representation that is particularly appealing for problems with long series with recurring patterns. The downside to using shapelets is the time complexity. The heuristic techniques described in recent research [4,5] offer potential speed up (often at the cost of extra memory) but are essentially different algorithms that are only really analogous to shapelets described in the original research [1]. Our interest is in optimizing the original shapelet finding algorithm within the context of the shapelet transform. We describe incremental improvements to the shapelet transform specifically for multi-class problems. Searching for shapelets assessed on how well they find a single class is more intuitive, faster and becomes more accurate than the alternative as the number of classes increases. We demonstrated that the binary shapelet approach is significantly more accurate than other shapelet approaches and is significantly more accurate than conventional approaches to the TSC problem.

References

1. Ye, L., Keogh, E.: Time series shapelets: a novel technique that allows accurate, interpretable and fast classification. Data Min. Knowl. Disc. **22**(1–2), 149–182 (2011)
2. Hills, J., Lines, J., Baranauskas, E., Mapp, J., Bagnall, A.: Classification of time series by shapelet transformation. Data Min. Knowl. Disc. **28**(4), 851–881 (2014)
3. Mueen, A., Keogh, E., Young, N.: Logical-shapelets: an expressive primitive for time series classification. In: Proceeding 17th ACM SIGKDD International Conference on Knowledge Discovery and Data Mining (2011)
4. Rakthanmanon, T., Keogh, E.: Fast-shapelets: a fast algorithm for discovering robust time series shapelets. In: Proceeding 13th SIAM International Conference on Data Mining (SDM) (2013)
5. Grabocka, J., Schilling, N., Wistuba, M., Schmidt-Thieme, L.: Learning time-series shapelets. In: Proceeding 20th ACM SIGKDD International Conference on Knowledge Discovery and Data Mining (2014)
6. Gordon, D., Hendler, D., Rokach, L.: Fast randomized model generation for shapelet-based time series classification. arXiv preprint arXiv:1209.5038 (2012)
7. Lines, J., Bagnall, A.: Alternative quality measures for time series shapelets. In: Yin, H., Costa, J.A.F., Barreto, G. (eds.) IDEAL 2012. LNCS, vol. 7435, pp. 475–483. Springer, Heidelberg (2012). doi:10.1007/978-3-642-32639-4_58
8. Hills, J.: Mining time-series data using discriminative subsequences. PhD thesis, School of Computing Sciences, University of East Anglia (2015)
9. Rakthanmanon, T., Bilson, J., Campana, L., Mueen, A., Batista, G., Westover, B., Zhu, Q., Zakaria, J., Keogh, E.: Addressing big data time series: mining trillions of time series subsequences under dynamic time warping. ACM Trans. Knowl. Disc. Data, **7**(3) (2013)
10. Hall, M., Frank, E., Holmes, G., Pfahringer, B., Reutemann, P., Witten, I.: The WEKA data mining software: an update. SIGKDD Explor. **11**(1), 10–18 (2009)

11. Quinlan, J.R., et al.: Bagging, boosting, and c4.5. In: AAAI/IAAI, vol. 1, pp. 725–730 (1996)
12. Cortes, C., Vapnik, V.: Support-vector networks. Mach. Learn. **20**(3), 273–297 (1995)
13. Breiman, L.: Random forests. Mach. Learn. **45**(1), 5–32 (2001)
14. Rodriguez, J.J., Kuncheva, L.I., Alonso, C.J.: Rotation forest: a new classifier ensemble method. IEEE Trans. Pattern Anal. Mach. Intell. **28**(10), 1619–1630 (2006)
15. Lines, J., Davis, L., Hills, J., Bagnall, A.: A shapelet transform for time series classification. In: Proceeding of the 18th ACM SIGKDD International Conference on Knowledge Discovery and Data Mining (2012)
16. Lin, J., Keogh, E., Li, W., Lonardi, S.: Experiencing SAX: a novel symbolic representation of time series. Data Min. Knowl. Disc. **15**(2), 107–144 (2007)
17. Chen, Y., Keogh, E., Hu, B., Begum, N., Bagnall, A., Mueen, A., Batista, G.: The UCR time series classification archive (2015). http://www.cs.ucr.edu/~eamonn/time_series_data/
18. Bagnall, A., Lines, J., Bostrom, A., Keogh, E.: The UCR/UEA TSC archive. http://timeseriesclassification.com
19. Bagnall, A., Bostrom, A., Lines, J.: The UEA TSC codebase. https://bitbucket.org/TonyBagnall/time-series-classification

DAAR: A Discrimination-Aware Association Rule Classifier for Decision Support

Ling Luo[1,2(⊠)], Wei Liu[3], Irena Koprinska[1], and Fang Chen[2]

[1] School of Information Technologies, University of Sydney, Sydney, Australia
{ling.luo,irena.koprinska}@sydney.edu.au
[2] Data 61, CSIRO, Eveleigh, Australia
fang.chen@data61.csiro.au
[3] Faculty of Engineering and IT, University of Technology, Sydney, Australia
wei.liu@uts.edu.au

Abstract. Undesirable correlations between sensitive attributes (such as race, gender or personal status) and the class label (such as recruitment decision and approval of credit card), may lead to biased decision in data analytics. In this paper, we investigate how to build discrimination-aware models even when the available training set is intrinsically discriminating based on the sensitive attributes. We propose a new classification method called Discrimination-Aware Association Rule classifier (DAAR), which integrates a new discrimination-aware measure and an association rule mining algorithm. We evaluate the performance of DAAR on three real datasets from different domains and compare DAAR with two non-discrimination-aware classifiers (a standard association rule classification algorithm and the state-of-the-art association rule algorithm SPARCCC), and also with a recently proposed discrimination-aware decision tree method. Our comprehensive evaluation is based on three measures: predictive accuracy, discrimination score and inclusion score. The results show that DAAR is able to effectively filter out the discriminatory rules and decrease the discrimination severity on all datasets with insignificant impact on the predictive accuracy. We also find that DAAR generates a small set of rules that are easy to understand and applied by users, to help them make discrimination-free decisions.

Keywords: Discrimination-aware data mining · Association rule classification · Unbiased decision making

1 Introduction

The rapid advances in data storage and data mining have facilitated the collection of a large amount of data and its use for decision making in various applications. Although applying data mining for automated or semi-automated decision making has many benefits, it also poses ethical and legal risks such as discrimination and invasion of privacy for the users, especially for socially sensitive tasks such as approving loans and credit cards, giving access to social benefits, assessing CVs of job applicants and hiring people. This paper focuses on building discrimination-aware classification models to eliminate the potential bias against sensitive attributes such as gender, age and ethnicity.

© Springer-Verlag GmbH Germany 2017
A. Hameurlain et al. (Eds.): TLDKS XXXII, LNCS 10420, pp. 47–68, 2017.
DOI: 10.1007/978-3-662-55608-5_3

Discrimination refers to the prejudicial treatment of individuals based on their actual or perceived affiliation to a group or class. People in the discriminated group are unfairly excluded from benefits or opportunities such as employment, salary or education, which are open to other groups [1]. In order to reduce the unfair treatment, there are anti-discrimination legislations in different countries such as the Equal Pay Act of 1963 and the Fair Housing Act of 1968 in the US; the Sex Discrimination Act 1975 in the UK [1]; the Racial Discrimination Act of 1975, the Sex Discrimination Act of 1984, the Equal Opportunity Act 1984 and the Fair Work Act 2009 in Australia [2]. Therefore, it is imperative to consider eliminating discrimination in applications such as decision support systems, otherwise the companies might be sued or penalized for acting against the law.

1.1   Problem Definition

Our problem can be formally stated as follows. Suppose we are given a labeled dataset D with N instances, m nominal attributes $\{A_1, A_2, ..., A_m\}$, from which the attribute $S = \{s_1, s_2, ..., s_p\}$ has been identified as a sensitive attribute (e.g. race, gender, etc.), and a class attribute C. D is a *discriminatory dataset*, if there is an undesirable correlation between the sensitive attribute S and the class attribute C. For example, when performing credit history checks, if the probability P(*credit history = good|race = white*) is much higher than P(*credit history = good|race = black*), it is said that this dataset is biased against black people. The *discrimination severity of a classifier* is measured from two viewpoints by: (1) *Discrimination Score* (DS) (see Sect. 3.2) and (2) *Inclusion Score* (IncS) (see Sect. 3.3). These two scores are computed on the testing set using the predicted class labels, and they can handle both binary and multi-value nominal sensitive attribute S. The goal is to learn a classifier with low discrimination severity with respect to S, with minimal impact on the classification accuracy.

1.2   Motivating Examples

As an example, assume that we are designing a recruitment system for a company to predict if a new candidate is suitable for a job or not. If the historical data contains more males than females (as shown in Fig. 1), the prediction model may tend to favor the attribute gender. A prediction rule using gender or another sensitive attribute like marital status, may achieve high accuracy, but it is not acceptable as it is discriminating, which is both unethical and against the law. Sensitive attributes such as gender, race and religion should be taken as an information carrier of a dataset, instead of distinguishing factors [3]. Females may be less suitable for a given job as on average they might have less work experience or lower educational level. It is acceptable to use work experience and educational level in the prediction model.

Another motivating example is the Census Income dataset from the UCI Machine Learning repository [4], used in this paper. It contains information about 1200 people, described with 40 attributes (e.g. employment status, sex, region of residence and family members), and a class attribute assessing whether each person has high

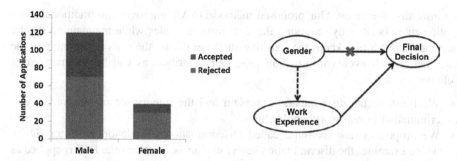

Fig. 1. Left:motivation example: the histogram of data in a recruitment system. Right: the prediction model using gender to predict the final decision may have higher accuracy but it is discriminating. Acceptable attributes such as work experience should be used.

Table 1. Distribution of class values (*high* or *low*) with respect to race

Values of *race*	Subtotal	High	Low	Ratio (High/Low)
White	521	433	88	4.920
Black	372	33	339	0.097
Others	307	134	173	0.775
Total	1200	600	600	1

(the income is greater than \$50 k per year) or low income. We explored the distribution of the class values with respect to the sensitive attribute *race*, to determine if there are any biased correlations between them. The results are shown in Table 1.

This dataset contains 600 people with high income and 600 people with low income and we can use the overall ratio between these two class values (i.e., 1) as a benchmark. If for a certain group this ratio is lower than 1 (i.e. this group has fewer *high* income than the benchmark), this means that the assessment is biased against this group; otherwise, the assessment favors this group. For example, we can see that for the group race = *white*, this ratio (4.92) is significantly higher than the benchmark, whereas for race = *black*, it is much lower than 1. This indicates that there is an undesirable correlation between race and assessment result in this dataset, which could lead to biased and unacceptable classification rules.

Our third motivating example is from the area of education. Suppose that the goal is to generate a rule that predicts the exam mark of the current students. If in the historic dataset used for training of the prediction algorithm, males have achieved significantly higher exam marks than females, a prediction rule using the attribute gender may be generated. It may produce high accuracy but we cannot use it for providing feedback to students or other decision making, as it can be seen as discriminating based on gender.

1.3   Contributions

In this paper we investigate discrimination-aware classification that aims to decrease the discrimination severity for sensitive attributes, when the training data contains intrinsic discrimination. Our proposed method DAAR improves the traditional association rule classifier by removing the discriminatory rules while maintaining similar accuracy. DAAR also keeps the sensitive attributes during the classifier training of the classifier, which avoids information loss. Our contributions can be summarized as follows:

- We illustrate the discrimination problem and the importance of minimizing discrimination in real world applications.
- We propose a new measure, called Discrimination Correlation Indicator (DCI), which examines the discrimination severity of an association rule. DCI is applied as an effective criterion to rank and select useful rules in discrimination-aware association rule classification tasks.
- We extend the standard definition of the Discrimination Score measure (DS) from binary to multi-level nominal sensitive attributes.
- We propose a new evaluation measure of discrimination severity, called Inclusion Score (IncS), for both binary and multi-level nominal sensitive attributes. Compared to DS, it provides a different viewpoint for evaluating the discrimination severity of a classifier.
- We propose DAAR, a new Discrimination-Aware Association Rule classification algorithm. We evaluate DAAR on three real datasets from different domains: traffic incident management, assessment of credit card holders and census income. We compare its performance with three methods: the standard association rule classifier [5], the state-of-the-art association rule classifier SPARCCC [6] and a discrimination-aware decision tree [7].

2   Related Work

2.1   Discrimination-Aware Methods

The discrimination-aware classification problem was introduced by Kamiran and Calders [8] and Pedreshi et al. [1], who formulated the direct and indirect discrimination definitions, and raised the attention of the Data Mining community to this problem. The existing discrimination-aware methods can be classified into two groups: methods that modify the dataset and methods that modify the algorithm.

The first group focuses on modifying the dataset during the pre-processing phase to eliminate the discrimination at the beginning. This includes removing the sensitive attribute, resampling [9] or relabeling some instances in the dataset to balance class labels for a certain sensitive attribute value [8, 10]. These methods typically lead to loss of important and useful information and undermine the quality of the predictive model that is learnt from the modified dataset. Additionally, just removing the sensitive attribute doesn't help due to the so called red-lining effect – the prediction model will still discriminate indirectly through other attributes that are highly correlated with the sensitive attribute [1, 11, 12].

The second group includes methods that integrate discrimination-aware mechanisms when building the classifier. Previous work [3, 7, 13, 14] have adapted various widely used classification algorithms, including decision trees, naïve Bayes, logistic regression and support vector machines to deal with potential discrimination issues.

The Discrimination-Aware Decision Tree (DADT) [7] applies two techniques to reduce discrimination: discrimination-aware tree construction and leaf relabeling. The first technique involves using a different criterion for selecting the best attribute during the decision tree building by considering not only the accuracy of the split but also the discrimination that it causes. Specifically, a new measure called Information Gain relative to the Sensitive attribute (IGS) is introduced, and it is used together with the standard Information Gain which is relative to the Class (IGC). Three different combinations are considered: IGC-IGS, IGC/IGS and IGC+IGS. The second technique is applied after generating the preliminary tree – some of its leaves are relabeled to decrease the discrimination severity to less than a pre-defined non-discriminatory constraint $\varepsilon \in [0, 1]$ while losing as little accuracy as possible. The results showed that DADT was able to maintain high accuracy while lowering the discrimination score.

Calders and Verwer [3] proposed three methods for modifying the naïve Bayes algorithm to reduce discrimination: (1) modifying the probability $P(S|C)$ in such a way that the predictions become discrimination-free, (2) training a separate model for each attribute value of S (2NB model) and (3) adding a latent variable which represents the discrimination-free class label and optimizing the model parameters using expectation maximization. The methods were evaluated on both artificial and real-life datasets and the results showed the 2NB method performed best.

Kamishima [14] introduced a regularization term in the formulation of logistic regression to penalize discrimination. The proposed method was able to reduce discrimination in exchange of slightly lower accuracy. In terms of trade-off between accuracy and discrimination removal, it was less efficient than the 2NB method [3].

Ristanovski et al. [13] investigated discrimination-aware classification for imbalanced datasets. To reduce discrimination, they applied a similar approach to [14]. Specifically, they modified the standard empirical loss function that is minimized in support vector machines by adding a discrimination-aware loss term. They showed promising results especially for imbalanced datasets where the discriminated group is more present. The proposed method can be applied to both balanced and imbalanced datasets.

2.2 Association Rule Methods

An association rule takes the form $X \rightarrow Y$, where X and Y are disjoint item sets. X contains the set of antecedents of the rule, and Y is the consequent of the rule [15]. Given a dataset containing N instances and an association rule $X \rightarrow Y$, the support and confidence of this rule are defined as follows:

$$\text{Support } (X \rightarrow Y) = \sigma(X \cup Y)/N, \text{Confidence } (X \rightarrow Y) = \sigma(X \cup Y)/\sigma(X) \quad (1)$$

where $\sigma(\cdot)$ is the frequency of an item set $(\cdot)$. When learning association rules, we are interested in rules with high support and high confidence.

Firstly introduced in [5], Classification Based on Association rules (CBA) uses association analysis to solve classification problems. In CBA, only class attributes can appear in Y. When classifying a new instance, if there are multiple matching rules, the rule with the highest confidence will be used to determine the class label. This method will also be referred as "standard AR" in the rest of the paper. Our proposed method is based on CBA, as the rule-based classifier can produce easy-to-interpret models.

A number of sophisticated associative classifiers were proposed in recent years, with various techniques for rule discovery, ranking, pruning, and prediction. For example, CMAR [16] extends CBA by using the more efficient FP-growth algorithm for pattern mining and weighted χ^2 analysis; CPAR improves the efficiency by using greedy algorithm and dynamic programming [17]; CCCS [18] deals with classification of imbalanced datasets by using a new measure complement class support.

SPARCCC is a relative new variation of CBA, which adds a statistical test to discover rules positively associated with the class in imbalanced datasets [6]. SPARCCC introduced the use of p-value and Class Correlation Ratio (CCR) in the rule pruning and ranking. CCR is defined as:

$$CCR(X \rightarrow y) = corr(X \rightarrow y)/corr(X \rightarrow \neg y)$$
$$corr(X \rightarrow y) = (\sigma(X \cup y) * N)/(\sigma(X) * \sigma(y)) \tag{2}$$

where $\sigma(\cdot)$ is the frequency of an item set $(\cdot)$. The method retains rules with *corr (X → y) > 1* and *CCR > 1*, which condition guarantees that they are statistically significant in the positive associative direction $X \rightarrow y$, rather than in the opposite direction $X \rightarrow \neg y$. SPARCCC has been shown to significantly outperform CCCS [6].

To track changes and detect concept drifts over time, CAEP, an associative classifier integrating emerging patterns was introduced in [19]. The emerging patterns can be identified by checking if the supports of a rule changed significantly on different datasets [20]. An efficient algorithm of mining emerging patterns was proposed in [21], which constructed tree-based structure to find jumping emerging patterns, whose support increases significantly from 0. The emerging pattern mining has been applied in areas such as cancer diagnostics [22] and bioinformatics [23].

3　The Proposed Method DAAR

Our proposed method DAAR uses the new measure Discrimination Correlation Indicator (DCI), together with confidence and support, to efficiently select representative and non-discriminatory rules that can be used to classify new instances.

DAAR offers the following advantages: (1) unlike naïve methods which simply remove the sensitive attribute to deal with discrimination, DAAR keeps the sensitive attribute in the model construction to avoid losing useful information; (2) the new measure DCI is easy to compute and capable of filtering out discriminatory rules with minimal impact on the predictive accuracy; (3) DCI is flexible to use, which can be integrated in other algorithms, such as methods dealing with emerging patterns, to generate non-discriminative results; (4) DAAR generates a smaller set of rules than the standard AR and these discrimination-free rules are easy to use by the users.

3.1   DCI Measure

DCI is designed to measure the degree of discrimination for *each rule*. Given the rule $X \to y$, DCI is defined as

$$
\text{DCI} = \begin{cases} \dfrac{\left| P(C=y|\ S=S_{\text{rule}}) - P(C=y|\ S=S_{\text{others}}) \right|}{(P(C=y|\ S=S_{\text{rule}}) + P(C=y|\ S=S_{\text{others}}))} \\ 0 \qquad\qquad\quad \text{if either of the above } P(\cdot) \text{ is } 0 \end{cases} \tag{3}
$$

$P(C=y|S=S_{\text{rule}})$ is the probability of the class to be y given the value of the sensitive attribute S is S_{rule}. If S is a binary attribute, S_{rule} is the value of S in the target rule and S_{others} is the other value of S. It is worth noting that, in general cases, the conditional probabilities $P(C=y|\ S=S_{\text{rule}})$ and $P(C=y|\ S=S_{\text{others}})$ do not add up to 1, since they are based on different conditions and they should be distinguished from $P(S=S_{\text{rule}}|C=y)$ and $P(S=S_{\text{others}}|C=y)$. If S is a multi-value nominal attribute, S_{others} includes the set of all attribute values except the one which appears in the target rule.

For example, if the target rule is "gender = female, housing = rent $\to$ assessment = bad", where gender is the sensitive attribute, then S_{rule} is female and S_{others} is male. The DCI for this rule will be:

$$
\text{DCI} = \frac{\left| P(C=\text{low}|\text{gender}=\text{female}) - P(C=\text{low}|\text{gender}=\text{male}) \right|}{P(C=\text{low}|\text{gender}=\text{female}) + P(C=\text{low}|\text{gender}=\text{male}\ )}
$$

If the sensitive attribute does not appear in that rule at all, we define DCI to be 0. Therefore, the range of DCI is [0, 1). When DCI equals to 0, which means the probability of the class value to be y is the same given different sensitive attribute values, the rule is considered to be free of discrimination. Otherwise, DCI is *monotonically increasing* with the discriminatory severity of a rule, which means that the larger DCI is, the more discriminatory the rule is with regard to the sensitive attribute S. When multiple rules have the same sensitive attribute S_{rule}, their DCI values only depend on the class label of these rules. This means that if the rules also have the same class label, they will get the same DCI, regardless the values of other attributes.

3.2   Discrimination Score

The Discrimination Score measure (DS) has been used in previous research [3, 7, 13] to *evaluate the discrimination severity of a classifier*. The conventional definition is only for the binary sensitive attribute case. If the sensitive attribute S is binary with values S_1 and S_2, DS is defined as:

$$
\text{DS} = \left| P(C=C_{\text{target}}|S=S_1) - P(C=C_{\text{target}}|S=S_2) \right| \tag{4}
$$

DS computes the difference between the probabilities of the target class C_{target} given the two values of the sensitive attribute $S=S_1$ or $S=S_2$, on the testing dataset. C_{target} can be any attribute value of the class label.

We extend this definition for the case with multi-value nominal attribute with m values, $m > 2$. We propose that DS is computed for each S_{value} and then averaged over the m scores. For each computation, it takes S_{value} as S_i and all the other values as S_{others}, and is defined as follows:

$$DS = 1/m * \left(\sum_{i=1}^{m} \left| P(C = C_{target} | S = S_i) - P(C = C_{target} | S = S_{others}) \right| \right) \quad (5)$$

The best case is when DS is zero, which means that the probabilities of the class value to be C_{target}, for all different values of the sensitive attribute, are the same, i.e. there is no discrimination. Otherwise, higher DS corresponds to higher discrimination severity. As the testing dataset has been labeled by the classifier, higher discrimination in the dataset indicates the classifier is biased, which should be prevented.

The purpose of DS and DCI is different. We note that DS cannot be used to filter discriminatory rules in DAAR instead of DCI, as DS is not applicable for a single rule as required in DAAR. More specifically, DS is designed to measure the quality of a classifier (any classifier, not only AR) based on a testing dataset that has been labeled by the classifier. In contrast, DCI is computed for each rule (hence, it requires a rule-based classifier) and then compared against a threshold to check whether the rule is discriminatory or not. Another difference between DCI and DS is that DCI is a single ratio, as there is only one possible attribute value of S in one rule, so S_{rule} and S_{others} are fixed once we know the target rule. On the other hand, DS is the average score over m sub-scores, as it computes a sub score for each possible attribute value S_{value} in the dataset, which will be more than one for a non-binary attribute S.

3.3 Inclusion Score

In addition to DS, we propose a new measure, called Inclusion Score (IncS), to *evaluate the discrimination severity of a classifier* from a different perspective. IncS measures whether the distribution of the values of the sensitive attribute S are uniform given the target class C_{target}. The relation between IncS and DS is similar to the relation between precision and recall. For example, assume that we want to assign tasks to employees and our aim is to avoid assigning difficult tasks to the same group of employees. IncS measures, among the difficult tasks, how uniformly are these tasks assigned to different groups of employees. In contrast, DS measures how uniform are the proportion of employees in each group, who have been assigned the difficult tasks.

The definition of IncS can also be split into two cases based on the number of sensitive attribute values: binary and multi-value.

If the sensitive attribute S is a binary nominal attribute with values S_1 and S_2, IncS is defined as the difference between the probabilities of the sensitive attribute $S = S_1$ or $S = S_2$ given C_{target} on the testing dataset

$$IncS = \left| P(S = S_1 | C = C_{target}) - P(S = S_2 | C = C_{target}) \right| \quad (6)$$

If the sensitive attribute S is a multi-value nominal attribute with m values, $m > 2$, we first compute sub-IncS for each S_{value} and then the overall IncS is the standard deviation of all m scores. For each computation, the current attribute value S_{value} is S_i and all the other values are S_{others}. The equations are as follows:

$$\text{IncS}_i = \left| P(S = S_i | C = C_{target}) - P(S = S_{others} | C = C_{target}) \right|$$

$$\mu = \frac{1}{m} * \left(\sum_{i=1}^{m} \text{IncS}_i \right), \text{IncS} = \sqrt{\frac{1}{m} * \left(\sum_{i=1}^{m} (\text{IncS}_i - \mu)^2 \right)} \tag{7}$$

We use the standard deviation of m values instead of the mean since both conditional probability terms in Eq. 7 have the same denominator. In most cases, $P(S = S_i | C = C_{target}) - P(S = S_{others} | C = C_{target})$ is negative for S_i, and the mean of absolute difference μ would be $(m - 2)/m$, which is only determined by the number of sensitive attribute values. Therefore, the standard deviation is more appropriate to measure whether the sub-IncS are uniform or diverse.

IncS is zero when there is no discrimination. In this case the probabilities of the different values of the sensitive attribute for class C_{target} will be the same. Higher IncS values mean larger differences among all $P(S = S_i | C = C_{target})$. More specifically, higher IncS shows the entities classified as C_{target} are dominated by some values of the sensitive attribute.

IncS can be used together with DS to evaluate the discrimination severity from different viewpoints. The classifier is desired to have high accuracy, low DS and low IncS. Sometimes, it is impossible to have low values for both DS and IncS, since the correlation between the two scores depends on the distribution of the values of the sensitive attribute. For example, assume that there are two groups, one with 200 people, and another one with 180 people. If both of them have 25% positive labels, DS will be 0, and IncS will be 0.053 (computed as $|200 * 25\% - 180 * 25\%|/380 * 25\%$), i.e. both scores are small. However, if the second group has 40 people, the IncS will be 0.367 due to the large difference between group sizes, while DS will be still 0. Therefore, for datasets with imbalanced sensitive attribute values, there is a trade-off decision between DS and IncS. If it is more important to have a similar proportion of a class value (e.g. difficult tasks) for each group, methods with lower DS should be preferred. If it is desirable to have a similar number of entities from each group, despite the different group sizes, methods with lower IncS should be preferred.

3.4   DAAR Algorithm

DAAR integrates DCI and the association rule classification to select discrimination-aware rules from all rules that have passed the minimum confidence and support thresholds. DAAR's algorithm is shown in Table 2.

The algorithm defines the maximum length of the rule as k in the input, so as a result, all rules will contain at most $k - 1$ antecedents on the left and one class label on the right. In the loop, the algorithm merges the $(i-1)$-item rule set which was generated

Table 2. DAAR algorithm

Algorithm Build Discrimination-Aware Classifier	
Input	dataset D with sensitive attribute S;
	$max\_length = k$; thresholds *conf*, *spt* and *dci*;
Output	non-discriminatory rules
1	**for** i = 2 to k // generate *i*-item rules
2	**if** i = 2: generate 2-item rule, which is the base case;
3	**else** : merge (i-1)-item RuleSet and 2-item RuleSet;
4	prune rules to keep anti-monotonic;
5	**end if**
6	set up contingency tables to calculate confidence, support, DCI;
7	filter rules using thresholds *conf*, *spt* and *dci*;
8	store rules in *i*-item RuleSet;
9	**end for**
10	sort rules in k-item RuleSet by DCI in ascending order;
11	**return** k-item RuleSet as the classifier;

in the last round with the 2-item rule set (the base case), to get the *i*-item rule set. In line 10, the set of rules is sorted by DCI in ascending order for clear presentation to users, and this sorting does not affect the classification results. The majority voting is then used to classify new instances; if the vote is tied (e.g. the same numbers of rules support each class), the sum of DCI of all rules supporting each class is calculated and compared to determine the final class. As discussed in Sect. 3.1, the severity of discrimination is lower when DCI is smaller; therefore the voting will select the class value with lower sum value as it is less discriminatory.

To illustrate this with an example, let's apply the standard AR and DAAR to our traffic incident dataset, in which the manager is the sensitive attribute. Two of the rules generated by the standard AR are "location = Cahill Expressway, Sydney → easy incident" and "manager = Henry → easy incident". Both of these rules have confidence of 0.76, but the second rule includes the sensitive attribute, and its DCI is 0.259. If the threshold for DCI is set as 0.1, our method can filter out the second rule effectively. More examples will be presented in Sect. 5.1.

When applying the non-discriminatory rules on new data (e.g., test datasets), the accuracy, DS and IncS are computed to evaluate the model.

4   Datasets and Experimental Setup

To evaluate the performance of DAAR, we use three real datasets from different domains, such as public transport and finance management. For each dataset we also compute the discrimination score before applying the classifiers, in the same way as described in Sect. 3.2, except that this was done for the whole dataset not just for the test set. This discrimination score can reveal the *intrinsic* correlation between the sensitive attribute and the class label of the dataset, and can act as a baseline. If the intrinsic discrimination score is high, it means the undesirable correlation or discrimination in this dataset is strong. Ideally, the discrimination score on the test data should be lower than the intrinsic discrimination score, which means that the classification model could mitigate the discrimination existing in the data and can avoid further discrimination.

4.1 Datasets

Traffic Incident Data was collected by the road authority of a major Australian city[1]. Each instance has 10 attributes, including the time, location and severity of the incident, the manager and other useful information. The *incident manager* is selected as the sensitive attribute S, and the class label is the *duration of the incident*, which takes two values: *long* and *short*. Our task is to predict whether an incident would be difficult to manage based on the available information. The *incident duration* is considered as a proxy to incident difficulty level – an incident with long duration corresponds to a difficult-to-manage incident and an incident with short duration corresponds to an easy-to-manage incident. Our experimental evaluation tests whether the proposed method can reduce the discrimination based on the sensitive attribute *incident manager* in predicting the *difficulty level of the incident*.

The data was preprocessed in two steps. Firstly, we noticed that there were more than 90 distinct manager values appearing in the full dataset, but most of them were only associated to less than 10 incidents. For simplicity, the managers were sorted by the number of associated incidents, and only instances handled by the top 5 managers were used in the experiment. Then, the majority class in the dataset was under-sampled to keep the dataset balanced with respect to the class attribute. This resulted in a dataset of 4,920 incidents, half of which were *difficult* and the other half were *easy* to manage. The intrinsic discrimination score of the traffic data is 0.232.

German Credit Card Data is a public dataset from the UCI Machine Learning repository [4]. The dataset consists of 1,000 examples (700 *good* and 300 *bad* customers), described by 20 attributes (7 numerical and 13 categorical). The sensitive attribute is the *personal status and sex*, which shows the gender of a customer and whether he or she is single, married or divorced. Since it can be discriminatory to assess customers by their gender and marital status, we would like to decrease the discrimination based on this attribute when classifying new customers.

As the original dataset is strongly biased towards the class *good*, with ratio *good*: *bad* = 7:3, we randomly removed 400 good customers to keep the balance of the dataset. This resulted in 600 examples, 300 from each of the two classes. The intrinsic discrimination score of this dataset is 0.21.

Census Income Data is also a public dataset from the UCI Machine Learning repository. It contains 40 attributes (7 numerical and 33 categorical), which are used to predict the income level of a person. If the income is over $50 K, the person is

---

[1] Data Collected by NSW Live Traffic: https://www.livetraffic.com/desktop.html#dev.

classified as having *high income*, otherwise as having *low income*. The attribute *race* (with values: *white, black, asian or pacific islander, amer indian aleut or eskimo* and *other*) is the sensitive attribute. We randomly selected a smaller portion of the original dataset containing 1,200 examples; half with *high income* and half with *low income*. The intrinsic discrimination score of this dataset is 0.422, which is about twice of the intrinsic discrimination score of the other two datasets.

4.2 Experimental Setup

We compare the performance of DAAR (in terms of accuracy, discrimination score and inclusion score) with three other methods: CBA (the standard AR), SPARCCC and DADT. CBA was chosen as it is a standard association rule classifier. SPARCCC was selected as it is a state-of-the-art association rule classifier for imbalanced datasets. The discriminatory dataset can be considered as a special type of imbalanced dataset. In a discriminatory database, the bias is against a certain class label within a group, having the same value for a sensitive attribute, e.g. *race = black*, while in an imbalanced dataset the bias is against a class over the whole dataset. DADT was selected as it is a successful discrimination-aware classifier.

All three association rule mining methods (standard AR, DAAR and SPARCCC) use confidence and support thresholds to remove the uninteresting rules. These thresholds are controlled as a baseline, while the other measures, the CCR threshold in SPARCCC and the DCI threshold in DAAR, are varied to generate comparison conditions. For example, for the traffic data, we used the following pairs of confidence and support values (*conf* = 0.6, *spt* = 0.01; *conf* = 0.6, *spt* = 0.03; *conf* = 0.6, spt = 0.05; *conf* = 0.5, *spt* = 0.1).

The number of rules generated by the classifier is another important factor to consider. It affects both the accuracy and discrimination score, and is also very sensitive to the chosen thresholds for confidence, support, CCR and DCI. In order to compare the results fairly, it is important to make sure that the number of rules of comparable conditions are in the same range. Hence, once the thresholds for confidence and support are fixed for the standard AR, the thresholds for CCR and DCI (between [0, 1)) are configured such that SPARCCC and DAAR can generate 4–5 conditions where the number of rules is of the same order of magnitude.

DADT, the discrimination-aware decision tree [7], uses the addition of the accuracy gain and the discrimination gain, IGC+IGS, as a splitting criterion, and relabeling of some of the tree nodes to reduce the discrimination. The non-discriminatory constraint $\varepsilon \in [0, 1]$ is tuned to generate comparison conditions.

5 Results and Discussion

All reported accuracy, discrimination score and inclusion score are average values from 10-fold cross validation. The *p* value is the result of the independent two-sample *t*-test, which is used to statistically compare the differences in performance between DAAR and the other methods.

5.1   Traffic Incident Data

The sensitive attribute for this dataset is the *incident manager*, and the class label is the *incident difficulty level*. Table 3 shows the average predictive accuracy, discrimination and inclusion scores of the proposed DAAR method and the three methods used for comparison (standard AR, SPARCCC and DADT).

Table 3. Accuracy, discrimination score and inclusion score for the traffic incident data

Methods	Accuracy		Discrimination score		Inclusion score	
	Mean	Std	Mean	Std	Mean	Std
Standard AR	77.22%	0.021	0.236	0.010	0.193	0.006
SPARCCC	73.21%	0.059	0.266	0.064	0.222	0.061
DADT	74.19%	0.045	0.187	0.022	0.154	0.027
DAAR	76.32%	0.025	0.213	0.012	0.180	0.012

Table 3 shows that overall, considering both accuracy and discrimination, DAAR is the best performing algorithm – it has the second highest accuracy and the second lowest discrimination score, and also small standard deviations. DAAR is statistically significantly more accurate than SPARCCC (p = 0.041) and slightly more accurate than DADT (p > 0.05). Although AR is more accurate than DAAR, this difference is not statistically significant. In terms of discrimination score, DAAR has significantly lower discrimination score than both the standard AR (p = 0.009) and SPARCCC (p = 0.0016). DADT has the lowest discrimination score but its accuracy is impacted – it has the second lowest accuracy after SPARCCC. SPARCCC is the worst performing method – it has the lowest accuracy and highest discrimination score, and the largest variation for both measures.

The inclusion score results are similar to the discrimination score results, showing the same ranking of the algorithms. DAAR has the second lowest value, which is statistically significantly lower than SPARCCC (p = 0.007) and the standard AR (p = 0.009). Although the inclusion score of DADT is lower than that of DAAR, the difference is not statistically significant.

Figure 2 presents a scatter plot of the accuracy and discrimination score. Ideally, we would like to see points in the top-left corner of the figure, which corresponds to high accuracy and low discrimination score simultaneously. However, there is a trade-off between the two measures, as the filtering out of discriminating rules will normally lower the predictive accuracy. Given this trade-off, our aim is to select a method with lower discrimination score but no significant impact on accuracy.

The results in Fig. 2 are consistent with the results in Table 3. All individual results of DAAR (triangles) are clustered in the left part of the graph which corresponds to low discrimination score, and these scores are always lower than the results for the standard AR (diamonds). SPARCCC is more diverse – it has points with relatively low and very large discrimination score and some others with a large discrimination score, which explains the overall low discrimination score and the large standard deviation. DADT's

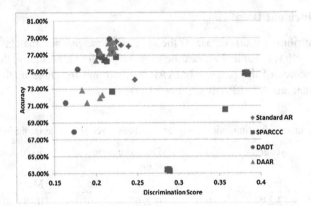

Fig. 2. Discrimination score and accuracy for the traffic incident data

results (circles) are also in the left part of the graph (low discrimination score) but the accuracy varies, which explains the overall lower accuracy and its higher standard deviation.

Figure 3 shows a scatter plot of the accuracy and inclusion score. Similarly to the trade-off between discrimination score and accuracy shown in Fig. 2, there is also a trade-off between the inclusion score and accuracy, which we can see in Fig. 3. As an exception, two of the points for DADT are in the bottom-left corner – they have both low accuracy and low inclusion score.

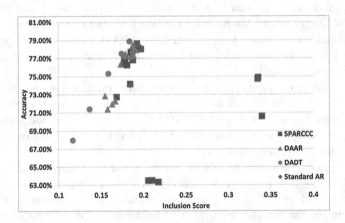

Fig. 3. Inclusion score and accuracy for the traffic incident data

Figure 4 presents a scatter plot of the discrimination score and inclusion score for this dataset. We can observe that the two scores are positively correlated. Desirable methods should be located in the bottom-left corner, where the values of both scores are low. This is the case for DADT (circles) and our method DAAR (triangles) – their points appear at the bottom-left part. In contrast, SPARCCC (squares) has several points with high discrimination and inclusion scores.

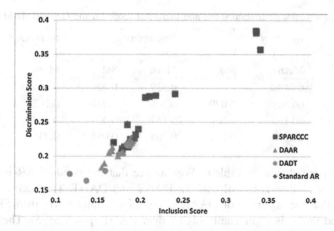

Fig. 4. Correlation between inclusion and discrimination score for the traffic incident data

Table 4 presents some of the rules generated by the DAAR, together with their confidence and DCI values. These rules are easy to understand by users, which is one of the advantages of applying association rule classification.

Table 4. Examples of rules generate by DAAR for the traffic incident data

Rules	Confidence	DCI
type = road development, time = [20:00,23:59] → difficult incident	1	0
severity = 2, time = [20:00,23:59], direction = north/south → difficult incident	0.95	0
day_of_week = 7, direction = north/south → difficult incident	0.90	0
location = Cahill Expressway, Sydney → easy incident	0.76	0

Table 5 shows more examples of rules that discriminate based on the manager with high confidence. These rules were filtered out by the proposed DAAR method.

Table 5. Examples of discriminating rules filtered out by DAAR

Rules	Confidence	DCI
manager = Frank, incident_severity = 2 → difficult incident	0.87	0.273
manager = Charles → difficult incident	0.75	0.245
manager = Henry → easy incident	0.76	0.259

5.2 German Credit Card Data

For this dataset, the aim is to eliminate the discrimination on personal status and sex when determining whether a customer is *good* or *bad*.

Table 6. Accuracy, discrimination score and inclusion score for the German credit card data

Methods	Accuracy		Discrimination score		Inclusion score	
	Mean	Std	Mean	Std	Mean	Std
Standard AR	68.40%	0.012	0.329	0.006	0.164	0.003
SPARCCC	66.60%	0.029	0.305	0.071	0.162	0.025
DADT	60.13%	0.008	0.157	0.007	0.262	0.009
DAAR	64.90%	0.027	0.208	0.055	0.231	0.035

The results are shown in Table 6. We can see that the standard AR is the most accurate method, followed by SPARCCC, DAAR and DADT. The statistical testing results show that the accuracy of DAAR is not significantly lower than SPARCCC ($p > 0.05$) and that it is significantly higher than DADT ($p = 5.5e{-}5$). The accuracy range on this dataset (60–68%) is lower compared to the traffic dataset (73–77%). This might be due to the small size of the credit card dataset and the large number of its attributes. In terms of discrimination score, DADT is the best performing algorithm, followed by DAAR, SPARCCC and AR. The t-test results show that our method DAAR has statistically significantly lower discrimination score than both the standard AR ($p = 8.0e{-}6$) and SPARCCC ($p = 0.0004$). Again, we can see that DAAR provides a good balance in terms of accuracy and discrimination score.

The inclusion scores range from 0.162 to 0.262. The results show that standard AR and SPARCCC have significantly lower inclusion scores than DAAR and DADT. Additionally, the inclusion score of DAAR is statistically significantly lower than DADT ($p = 0.012$). The performance ranking of the four methods is in reverse order compared to the ranking based on the discrimination score. Figure 5 demonstrates that the discrimination scores and the inclusion scores on this dataset are negatively correlated, so all points distribute along the secondary diagonal of the plot. Consistent with the inclusion score property discussed in Sect. 3.3, DAAR and DADT could not get low inclusion scores on this dataset due to the imbalanced distribution of the *personal status and sex* values.

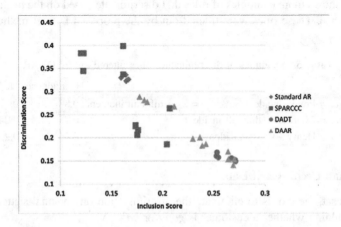

Fig. 5. Correlation between inclusion and discrimination score for the German credit card data

The scatter plot in Fig. 6 illustrates clearly the trade-off between accuracy and discrimination score. We can see that all DAAR's points (triangles) are on the left with respect to the standard AR points, but the accuracy of these points is lower than the accuracy of the standard AR points due to the removal of the discriminating rules. It is also interesting to observe that although SPARCCC has higher average accuracy than DAAR, the scatter plot demonstrates that for the same discrimination score (between 0.15 and 0.3), SPARCCC has lower accuracy than DAAR. The SPARCCC points are grouped into two main clusters: one in the middle that has similar discrimination score as DAAR and three points at the right corner that have high accuracy but large discrimination scores. DADT points are clustered at the bottom left of the graph, so that its accuracy lower than DAAR's for similar discrimination scores.

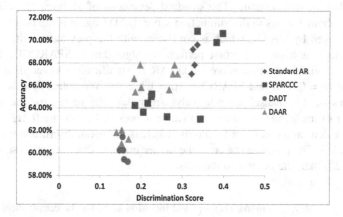

Fig. 6. Discrimination score and accuracy for the German credit card data

Since the results with low discrimination score have high inclusion score on this dataset (as shown in Fig. 5), we find that the DADT results (circles) cluster at the bottom-right corner in Fig. 7. DAAR performs better - most of the DAAR points

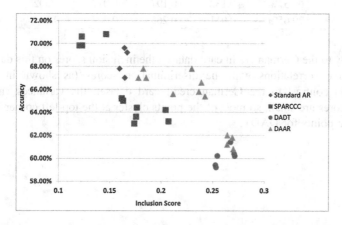

Fig. 7. Inclusion score and accuracy for the German credit card data

(triangles) have inclusion scores between 0.15 and 0.25, and higher accuracy than DADT.

5.3   Census Income Data

The sensitive attribute in this dataset is *race* and the class label is *income level*. The aim is to avoid predicting the income level (*high* or *low*) of a person based on their race.

From Table 7, we can see that in terms of average accuracy, all methods except DADT perform very similarly achieving accuracy of about 79–81%, which is higher than the accuracy on the previous two datasets. Therefore, most of points in the scatter plot (Fig. 8) cluster on the top-right corner. SPARCCC is slightly more accurate and DAAR is slightly less accurate. The standard deviations of all three association rule methods are about 1%. As to discrimination score, DADT again is the best performing method, followed by DAAR, which is consistent with the previous results. The standard AR comes next and the worst performing algorithm is SPARCCC. The t-test shows that the discrimination score of DAAR is significantly lower than both the standard AR ($p = 0.016$) and SPARCCC ($p = 1.12e{-}6$). We notice that the discrimination scores of this dataset are relatively larger than the scores for the other two datasets. For example, the mean discrimination scores of DAAR are 0.213 and 0.208 for traffic incident and credit card datasets, whereas the mean discrimination score of DAAR is 0.265. The possible reason is the higher intrinsic discrimination score of this dataset (0.422) than the other two datasets.

Table 7. Accuracy, discrimination score and inclusion score for the census income data

Methods	Accuracy		Discrimination score		Inclusion score	
	Mean	Std	Mean	Std	Mean	Std
Standard AR	80.81%	0.011	0.285	0.009	0.274	0.003
SPARCCC	81.20%	0.010	0.289	0.010	0.274	0.002
DADT	68.57%	0.132	0.197	0.077	0.302	0.012
DAAR	79.65%	0.011	0.265	0.007	0.280	0.003

Similarly to the German credit card dataset, the inclusion scores on this dataset also have negative correlations with the discrimination scores (as shown in Fig. 10). However, in contrast to the German credit card dataset, the variances among the inclusion scores are small, so most of the points cluster at the top-left corner in Fig. 9, except three points for DADT.

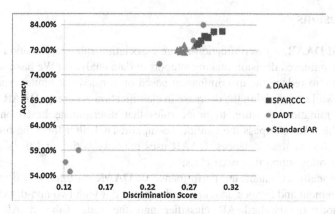

Fig. 8. Discrimination score and accuracy for the census income data

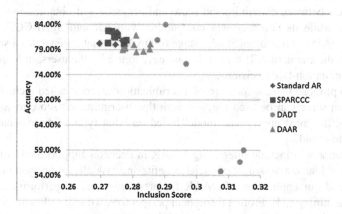

Fig. 9. Inclusion score and accuracy for the census income data

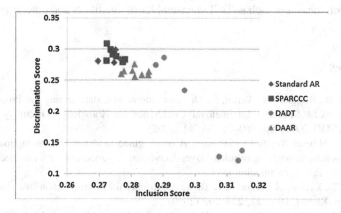

Fig. 10. Correlation between inclusion and discrimination score for the census income data

6 Conclusions

We proposed DAAR, a discrimination-aware association rule classification algorithm that provides unbiased decision making support in data analytics. We have shown that DAAR is able to reduce the discrimination based on sensitive attributes, such as race and gender, while minimizing the decrease in classification accuracy. DAAR uses DCI, a new discrimination measure, to prune rules that discriminate based on sensitive attributes. The rules that pass the confidence-support-DCI filter form the final DAAR rule set. To classify new instances, DAAR uses majority voting and the sum of DCI scores of the rules supporting each class.

We empirically evaluated the performance of DAAR on three real datasets from traffic management and finance domains, and compared it with two non-discrimination-aware methods (a standard AR classifier and the state-of-the-art AR classifier SPARCCC), and also with the discrimination-aware decision tree DADT. The experimental results on all datasets consistently showed that DAAR is capable of providing a good trade-off between discrimination score and accuracy – it obtained low discrimination score while its accuracy was comparable with AR and SPARCCC, and was higher than DADT. An additional advantage of DAAR is that it generates a smaller set of rules than the standard AR; these rules are easy to use by the users, in helping them make discrimination-free decisions.

We also proposed a new measure of discrimination severity called inclusion score. The inclusion score can be used together with the discrimination score to support the users to find the most appropriate model based on what type of discrimination they would like to avoid.

Future work will include integrating DAAR in decision support applications such as approval of loans, assessment of social benefits and task allocation system. We also plan to extend our application of DAAR for predicting student performance in educational data mining [24]. From a theoretical perspective, we will follow three research directions: (1) investigate the case with multiple sensitive attributes by considering the importance of each attribute, (2) combine the discrimination and inclusion scores into a single measure for evaluating the severity of the discrimination and (3) study the use of discrimination measures such as DCI in ensembles of classifiers and emerging patterns.

References

1. Pedreshi, D., Ruggieri, S., Turini, F.: Discrimination-aware data mining. In: Proceedings of the 14th ACM SIGKDD International Conference on Knowledge Discovery and Data Mining (KDD 2008), pp. 560–568. ACM (2008)
2. Australian Human Rights Commission, A quick guide to Australian discrimination laws. https://www.humanrights.gov.au/employers/good-practice-good-business-factsheets/quick-guide-australian-discrimination-laws
3. Calders, T., Verwer, S.: Three naive Bayes approaches for discrimination-free classification. Data Min. Knowl. Disc. **21**, 277–292 (2010)
4. UCI Machine Learning Repository, University of California, Irvine, School of Information and Computer Sciences. http://archive.ics.uci.edu/ml

5. Ma, Y., Liu, B., Yiming, W.H.: Integrating classification and association rule mining. In: Proceedings of the 4th ACM SIGKDD International Conference on Knowledge Discovery and Data Mining (KDD 1998), pp. 80–86 (1998)
6. Verhein, F., Chawla, S.: Using significant, positively associated and relatively class correlated rules for associative classification of imbalanced datasets. In: Proceedings of the 7th IEEE International Conference on Data Mining, pp. 679–684. IEEE (2007)
7. Kamiran, F., Calders, T., Pechenizkiy, M.: Discrimination aware decision tree learning. In: Proceedings of the 10th IEEE International Conference on Data Mining, pp. 869–874. IEEE (2010)
8. Kamiran, F., Calders, T.: Classifying without discriminating. In: International Conference on Computer, Control and Communication, pp. 1–6. IEEE (2009)
9. Kamiran, F., Calders, T.: Classification with no discrimination by preferential sampling. In: Proceedings of the Benelearn (2010)
10. Calders, T., Kamiran, F., Pechenizkiy, M.: Building classifiers with independency constraints. In: IEEE International Conference on Data Mining Workshops, pp. 13–18. IEEE (2009)
11. Hajian, S., Domingo-Ferrer, J.: A methodology for direct and indirect discrimination prevention in data mining. IEEE Trans. Knowl. Data Eng. 25, 1445–1459 (2013)
12. Pedreschi, D., Ruggieri, S., Turini, F.: Integrating induction and deduction for finding evidence of discrimination. In: Proceedings of the 12th International Conference on Artificial Intelligence and Law, pp. 157–166. ACM, Barcelona (2009)
13. Ristanoski, G., Liu, W., Bailey, J.: Discrimination aware classification for imbalanced datasets. In: Proceedings of the 22nd ACM International Conference on Information & Knowledge Management, pp. 1529–1532. ACM (2013)
14. Kamishima, T., Akaho, S., Asoh, H., Sakuma, J.: Fairness-aware classifier with prejudice remover regularizer. In: Joint European Conference on Machine Learning and Knowledge Discovery in Databases, pp. 35–50. Springer (2012)
15. Simon, G.J., Kumar, V., Li, P.W.: A simple statistical model and association rule filtering for classification. In: Proceedings of the 17th ACM SIGKDD International Conference on Knowledge Discovery and Data Mining, pp. 823–831. ACM, 2020550 (2011)
16. Li, W., Han, J., Pei, J.: CMAR: Accurate and efficient classification based on multiple class-association rules. In: Proceedings IEEE International Conference on Data Mining, pp. 369–376. IEEE (2001)
17. Yin, X., Han, J.: CPAR: Classification based on Predictive Association Rules. In: SDM, pp. 331–335. SIAM (2003)
18. Arunasalam, B., Chawla, S.: CCCS: a top-down associative classifier for imbalanced class distribution. In: Proceedings of the 12th ACM SIGKDD International Conference on Knowledge Discovery and Data Mining, pp. 517–522. ACM (2006)
19. Dong, G., Zhang, X., Wong, L., Li, J.: CAEP: classification by aggregating emerging patterns. In: Arikawa, S., Furukawa, K. (eds.) DS 1999. LNCS, vol. 1721, pp. 30–42. Springer, Heidelberg (1999). doi:10.1007/3-540-46846-3_4
20. Dong, G., Li, J.: Efficient mining of emerging patterns: discovering trends and differences. In: Proceedings of the Fifth ACM SIGKDD International Conference on Knowledge Discovery and Data Mining, pp. 43–52. ACM (1999)
21. Bailey, J., Manoukian, T., Ramamohanarao, K.: Fast algorithms for mining emerging patterns. In: Elomaa, T., Mannila, H., Toivonen, H. (eds.) PKDD 2002. LNCS, vol. 2431, pp. 39–50. Springer, Heidelberg (2002)

22. Li, J., Liu, H., Ng, S.-K., Wong, L.: Discovery of significant rules for classifying cancer diagnosis data. Bioinformatics **19**, ii93–ii102 (2003)
23. Li, J., Wong, L.: Emerging patterns and gene expression data. Genome Inform. **12**, 3–13 (2001)
24. Luo, L., Koprinska, I., Liu, W.: Discrimination-aware classifiers for student performance prediction. In: International Conference on Educational Data Mining (2015)

New Word Detection and Tagging on Chinese Twitter Stream

Yuzhi Liang[✉], Pengcheng Yin, and S.M. Yiu

Department of Computer Science,
The University of Hong Kong, Pokfulam, Hong Kong
{yzliang,pcyin,smyiu}@cs.hku.hk

Abstract. Twitter becomes one of the critical channels for disseminating up-to-date information. The volume of tweets can be huge. It is desirable to have an automatic system to analyze tweets. The obstacle is that Twitter users usually invent new words using non-standard rules that appear in a burst within a short period of time. Existing new word detection methods are not able to identify them effectively. Even if the new words can be identified, it is difficult to understand their meanings. In this paper, we focus on Chinese Twitter. There are no natural word delimiters in a sentence, which makes the problem more difficult. To solve the problem, we first introduce a method of detecting new words in Chinese twitter using a statistical approach without relying on training data for which the availability is limited. Then, we derive two tagging algorithms based on two aspects, namely word distance and word vector angle, to tag these new words using known words, which would provide a basis for subsequent automatic interpretation. We show the effectiveness of our algorithms using real data in twitter and although we focus on Chinese, the approach could be applied to other Kanji based languages.

Keywords: Chinese tweets · New word detection · Annotation · Tagging

1 Introduction

New social media such as Facebook or Twitter becomes one of the important channels for dissemination of information. Sometimes they can even provide more up-to-date and inclusive information than that of news articles. In China, Sina Microblog, also known as Chinese Twitter, dominates this field with more than 500 million registered users and 100 million tweets posted per day. An interesting phenomenon is that the vocabularies of Chinese tweets thesaurus have already exceeded traditional dictionary and is growing rapidly. From our observation, most of the new words are highly related to hot topics or social events, which makes them appear repeatedly in different social media and people's daily life. For example, the new word *"Yu'e Bao"* detected from our experimental dataset is an investment product offered through the Chinese e-commerce giant Alibaba. Its high interest rate attracted hot discussion soon after it first appeared, and without any concrete marketing strategy, *Yu'e Bao* has been adopted by 2.5

© Springer-Verlag GmbH Germany 2017
A. Hameurlain et al. (Eds.): TLDKS XXXII, LNCS 10420, pp. 69–90, 2017.
DOI: 10.1007/978-3-662-55608-5_4

million users who have collectively deposited RMB 6.601 billion ($1.07 billion) within only half a month.

Obviously, these *"Tweet-born"* new words in the Chinese setting are worthy of our attention. However, finding new words from Chinese tweet manually is unrealistic due to the huge amount of tweets posted every day. It is desirable to have an automatic system to analyze tweets. The obstacle is that Twitter users usually invent new words using non-standard rules that appear in a burst within a short period of time. Existing new word detection methods, such as [1,16], are not able to identify them effectively. Even if the new words can be identified, it is difficult to understand their meanings.

In this paper, we focus on Chinese Twitter. The contributions of our paper are listed as below:

- We introduce a Chinese new word detection[1] framework for tweets. This framework uses an unsupervised statistical approach without relying on hand-tagged training data for which the availability is very limited. The proposed framework is compared with ICTCLAS and Stanford Chinese-word-segmenter on new word detection over real microblog data (2013-07-31 to 2013-08-06). The result shows our method is competitive in new word detection regarding precision and recall rate. Although we focus on Chinese, the approach could be applied to other Kanji based languages like Japanese or Korean.
- We propose a novel method to annotate Chinese new words in microblog by automatic tagging. Context Distance and Context Cosine Similarity are derived for the similarity measurement. To the best of our knowledge, it is the first time automatic tagging is used in word interpretation. The new word tagging result accuracy is measured by checking the existence of the generated tag words in corresponding Baidu Entry (Baidu Entry is an online encyclopedia like Wikipedia. Some new words are recorded in Baidu Entry serval months after its first appearance). The average precision of tagging by Context Distance and that of Context Similarity are 52% and 79% respectively.

2   Related Works

2.1   New Word Detection in Chinese Tweets

Unlike English and other western languages, many Asian languages such as Chinese and Japanese do not delimit words by spaces. An important step to identify new words in Chinese is to segment a sentence into potential word candidates. Existing approaches to Chinese new word detection fall roughly into two categories: supervised (or semi-supervised) method and unsupervised method.

The core idea of supervised method is transferring the segmentation task to a tagging problem, each character is represented as a one-hot vector or an n-gram vector then by using a pre-trained classifier, a tag of the character will be generated to indicates the position of the character in a word (i.e. 'B' for beginning, 'M'

---

[1] Also known as Out-of-Vocabulary (OOV) detection.

for middle, 'E' for end, 'S' for single character as a word). Supervised method is popular in Chinese word segmentation. For example, [22] using a shallow (2 layers) neural network as a classifier to tag the characters, [1] using a discriminative undirected probabilistic graphical model Conditional Random Field in [23] to perform the classification. Moreover, both of the two most widely used Chinese word segmentation/new word detection tool Stanford Chinese-word-segmenter (based on Conditional Random Field CRF [21]) and ICTCLAS (based on Hierarchical Hidden Markov model HHMM [16]) are using supervised method. The problem is, precision of supervised method often relies on the quality of tagged training set. Unfortunately, as far as we know, the largest public hand-tagged training dataset of Chinese Microblog only consists of around 50,000 sentences, which is not sufficient to capture all features of Chinese new words in tweets. On the other hand, differ from other documents, microblog tweets are short, informal and have multivariate lexicons (words can form new words in various ways) which makes training sets of traditional documents is not so suitable for microblog crops. The difference between semi-supervised method and supervised method is that semi-supervised method (e.g. [7,9]) tries to derive statistical information from the training datasets such as Description Length Gain (DLG) [19] in which the best segmentation of a sentence is computed to maximize information gain. Although this information can be used to identify new words, the computation of training dataset features is time-consuming and the accuracy still relies on the quality of training datasets. Existing solutions [14,15] for identifying new words specially designed for Chinese Microblog word segment are also supervised machine learning methods. Thus, both suffer from the shortage of good training datasets.

Unsupervised method perform Chinese word segmentation by deriving a set of context rule or calculating some statistical information from the target data. From our study, we notice that contextual rule-based approach is not suitable for the task of detecting new words from Chinese tweets because new words emerged from Sina Microblog are rather informal and may not follow these rules while statistical method is a good solution for this problem since it can be purely data driven.

2.2 New Word Annotation

Existing approaches for the new entity (phrase or words) interpretation include name entity recognition (NER) [8] and using the online encyclopedia as a knowledge base [4]. NER seeks to locate and classifies name entity into names of persons, organizations, locations, expressions of time, quantities, monetary values, and percentages, etc. [2,3,9]. However, the new words we detect in Sina tweets are not limited in name entity. Some of them are new adjectives such as " 坚韧淡定 " (clam and tough). Even though NER can classify the new entity into different categories, the meaning of the new entity is still missing. Another popular approach is interpreting entities by linking them to Wikipedia. This is not applicable for annotating new emerging words because most of new words will not have a corresponding/related entry in any online encyclopedias within a short period of time right after the new word comes out.

3 Overview

In this paper, we describe an unsupervised Chinese new word detection technic in Sect. 4. There is no training set required by this method such that it is a good match for new word detection in Chinese tweets whose high-quality training set is unavailable. Then for new word interpretation, we propose a novel framework in Sect. 5 to realize new word annotation by automatic tag the new words with known words. Some notations in this paper is listed in Table 1.

Table 1. Notations

Symbol	Description
s	Character sequence
n	The number of characters in s
c_i	The ith character in s
w	Word
w_{new}	New word
w_{known}	Known word
t	Target time point
t_0	A time point before t
D	Tweet corpus
T	A tweet
Set_{tweet}	Unsegment tweets
k	Number of words in Set_{known}
Set_{new}	New word set
Set_{known}	Known word set
Set_{cand}	Candidate word set
$doc(w_1, w_2)$	Document made by tweets containing w_1 while w_1 and w_2 are excluded from the document
$doc(w)$	Document made by all the tweets containing w

4 New Word Detection

In new word detection, our target is to define an efficient model to detect OOV words from Chinese Twitter stream while avoiding using tagged datasets. We proposed a new word detection framework by computing the word probability of a given character sequence. This approach combines ideas from several unsupervised Chinese word segmentation methods, i.e. [6,12,13], as follows. In statistical Chinese segmentation methods, a common basic assumption is that a Chinese word should appear as a stable sequence in the corpus. Symmetrical

Conditional Probability (SCP) [12] can be used to measure the cohesiveness of a given character sequence. On the other hand, Branching Entropy (BE) [13] measures the extent of variance based on the idea that if a character sequence s is a valid word, it should appear in different contexts. We note that these two statistical approaches measure the possibility of s being a valid word from two perspectives. SCP captures the cohesiveness of the characters in s, while BE considers the outer variability. They can complement each other in achieving accuracy. To further reduce the noise, we use a word statistical feature Overlap Variety in [6].

4.1 Definition of New Word

New word of time t is defined as words appears at time t but not exists at time $t_0(t_0 < t)$. In our study, we are only interested in words whose frequencies are larger than a certain threshold for the following reasons: Firstly, low-frequency character sequences usually are meaningless character sequences. Secondly, even some of them are valid words, they are mainly people names just known by the posters or misspelled words which are not our target. Thirdly, we need to anno-tate the new words after they are detected, low-frequency character sequence have not enough relevant data for automatic tagging. In the process of new word detection in Chinese tweets, firstly, we need to get the word set at time t_0 and the word set at time t from the unsegmented tweets. Then for any word w extracted from the unsegmented tweets at t $Set_{tweet}(t)$, if w is not exist in $Set_{tweet}(t_0)$, w is regarded as a new word, otherwise w is a known word.

4.2 Word Extraction

The first step is to extract word segments from a set of unsegmented tweets. We have discussed in the introduction that the state-of-art supervised method is not suitable for our application due to the lack of training corpus. Instead of relying on training data, we propose an unsupervised approach for Chinese new word detection from tweets. Different statistical based Chinese word segmentation approaches are jointly used in the proposed method. Symmetrical Conditional Probability (SCP) [12] is a method which evaluates the cohesiveness of a char-acter sequence while Branching Entropy (BE) [13] measures the environment variance of a character sequence. These two approaches can complement each other in achieving accuracy. Moreover, Overlap Variety [6] is further used in reducing noise. For each character sequence in the set of unsegmented tweets with a length between two and four, a probability score will be calculated to indicate how likely the character sequence is a valid word. Technical detail will be introduced in the following parts.

Sequence Frequency. Sequence Frequency is an important noise filtering crite-ria in Chinese word segmentation. It is base on the assumption that if s is a valid word, it should appear repeatedly in Set_{tweet}. In our study, character sequences whose frequency lower than a threshold, $Thres_{freq}$, are filtered beforehand.

Symmetrical Conditional Probability. Symmetrical Conditional Probability (SCP) is a statistical criterion which measures the cohesiveness of a
given character sequence s by considering all the possible binary segmentations of s. It based on the assumption if s is a valid word, the substrings of s will mainly appear along with s. For example, given sentence
"氨基酸/是/构成/蛋白质/的/基本/单位" (Amino acids constitute the basic unit

of protein). The character sequence "氨基酸" (Amino acids) is a valid word,
its substrings (i.e. "氨基","酸","氨","基酸") should mainly co-occur with
"氨基酸".

Formally, let n denotes the length of s, c_i denotes the ith character in s, the
possibility of the given sequence appearing in the text, which is estimated by its
frequency, the SCP score of s is by Eq. 1.

$$SCP(s) = \frac{freq(s)^2}{\frac{1}{n-1}\sum_{i=1}^{n-1} freq(c_1, c_i)freq(c_{i+1}, c_n)} \tag{1}$$

The value of $SCP(s)$ is in range $(-\infty, 1]$. $SCP(s)$ is high when all the binary
segmentations of s mainly appear along with s.

Branching Entropy. Branching Entropy (BE) measures the extent of the variance of the contexts in which s appears. It is based on the idea that if s is a
valid word, it should appear in different contexts. Branching Entropy quantifies
such context variance by considering the variance of the preceding and following
character of s. Assume X is the set of the preceding characters of s, $P(s|x)$ is
the probability that x is followed by s, the formula is defined by Eq. 2.

$$P(s|x) = \frac{P(x + s)}{P(x)} \tag{2}$$

Here $P(x + s)$ is the frequency of character sequence $(x + s)$ and $P(x)$ is the
frequency of x.

The Left Branching Entropy of $P(x|s)$ is defined as Eq. 3:

$$BE_{left}(s) = -\sum_{x \in X} P(s|x) \log P(s|x) \tag{3}$$

The Right Branching Entropy, $BE_{right}(s)$, can be defined similarly by considering the characters following s. The overall Branching Entropy of sequence
s is defined by Eq. 4:

$$BE(s) = \min\{BE_{left}(s), BE_{right}(s)\} \tag{4}$$

The value of $BE(s)$ is in range $[0, \infty)$. $BE(s) = 0$ if the size of the preceding character set of s or that of the succeeding character set of s equals
to 1. In other words, the value of $BE(s)$ is 0 when the left boundary or the

right boundary of s occurs in only one environment in the given corpus. For a specific s, the value of $BE(s)$ is high if both of the size of the preceding character set and the succeeding character set is large which means s often happens in different context such that s is likely to be a valid word. For instance, given the character sequence "门把手(doorknob)" which is a valid word, we might find it in different contexts such as "门把手坏了(The doorknob is broken)", "要一个新的门把手(Need a new doorknob)", "或者把这个门把手修好 (Or repair this doorknob)", "这个门把手很漂亮(This doorknob is pretty)". In this case, there are three different preceding characters of "门把手" (sentence start, '的', '个')", as well as four different succeeding characters ('坏', sentence end, 修', '很). The Left Branching Entropy of "门把手" is $BE_{left}(r) = -(\frac{1}{4} \times log(\frac{1}{4}) + \frac{1}{4} \times log(\frac{1}{4}) + \frac{1}{2} \times log(\frac{1}{2})) = 1.5$, the Right Branching Entropy of "门把手(doorknob)" is $BE_{right}(r) = -(\frac{1}{4} \times log(\frac{1}{4}) \times 4) = 2$, the overall Branching Entropy takes the minimum of these two values, i.e.,1.5. On the other hand, for an invalid character sequence "门把" (a subsequence of the valid word "门把手"), the number of different preceding of "门把" is also three (sentence start, "的', '个'), but the number of succeeding of "门把" is just one ('个'), so $BE_{left}(r) = 1.5$ while $BE_{right}(r) = -(1 \times log(1)) = 0$, and the overall BE value of "门把" is 0.

Word Probability Score. In the process of finding valid word from the set of character sequences, the character sequences have extremely low BE score or SCP score will be abandoned and the character sequences whose BE score or SCP are larger than the corresponding threshold can be selected as valid word directly beforehand. Then an word probability score $P_{word}(s)$ is defined for the reset of character sequences to indicates how likely a character sequence s is a valid word. The score is calculated based on normalized BE and SCP of s. s is probably a valid word if $P_{word}(s)$ is high. The formula of $P_{word}(s)$ is as Eq. 5.

$$P_{word}(s) = w_{BE} \times BE'(s) + w_{SCP} \times SCP'(s) \tag{5}$$

$BE'(s)$ is the normalized BE score of s which is defined by max-min normalization of the BE values Eq. 6

$$BE'(s) = \frac{BE(s) - min_{BE}}{max_{BE} - min_{BE}} \tag{6}$$

The value of $BE'(s)$ is in range [0,1].

$SCP'(s)$ is the normalized the SCP score of s. Experimental result shows that SCP scores of the character sequences are not normally distributed, the commonly used normalization method max-min normalization and z-score normalization is not efficient in this case. We use a shift z-score mentioned in [16] which can provide an shift and scaling z-score to normalize the majority of the SCP values into range [0, 1]. The formula is as Eq. 7.

$$SCP'(s) = \frac{\frac{SCP(s) - \mu}{3\sigma} + 1}{2} \tag{7}$$

Though Eq. 5, we evaluate the probability of a given character sequence being a valid word by combining its outer variance and inner cohesiveness. Basically, a valid Chinese word should occur in different environments (high BE) and the characters in the word should often occur together (high SCP). Take the previous "门把手 (doorknob)" example in the Branching Entropy part. We have shown that BE(门把手) is high, and from the corpus, we can see that character sequences "门把", "把手", "门把手" also have high co-occurrence which makes the overall P_{word}(门把手) high. The w_{BE} and w_{SCP} in Eq. 5 are the weights of $BE'(s)$ and $SCP'(s)$ in calculating $P_{validword}(s)$. Intuitively, we will set w_{BE} and w_{SCP} with the same value(i.e. $w_{BE} = 0.5$ and $w_{SCP} = 0.5$). However, from our study, we find that the BE value is more useful in Chinese new word detection in tweets. The reason might be that some of the "*Tweet-born*" new words are compound of several known words. Take the valid word "余额宝" (Yu-E Bao) "as an example, it is a compound of other two valid words "余额(balance)" and "宝(treasure)", so the denominator of SCP(余额宝) which contains $freq$(余额) $\times$ $freq$(宝) will become large and make the overall SCP(余额宝) relative small. In other words, SCP(余额宝) cannot efficiently identify "余额宝 (Yu-E Bao)" as an valid word in this case. Based on the above observation, we set w_{BE} slightly higher than w_{SCP}, which are 0.6 and 0.4 respectively.

Noise Filtering. Although a set of valid word candidates can be achieved by setting a threshold on $P_{word}(s)$, substrings of valid words exist as noise from our observation. For example, "国媒体(country media)" is an invalid segmentation, it is a substring of word/phrase such as "美国媒体(American media)" or "中国媒体(Chinese media)". However, P_{word}(国媒体) is not low. That is because on one hand, "国媒体(country media)" often appears together such that SCP(国媒体) is high. On the other hand, the preceding of "国媒体(country media)" can be "中[国](China)", "美[国](American)", "外[国](foreign countries)", etc. while the succeeding of "国媒体(country media)" can be "宣称(claim)", "报道(report)" which makes BE(国媒体) high as well. In other words, the character "国" is a component of many different words which makes the character sequences "国媒体" occurs in different environments even it is not a valid word. The basic idea of filtering this kind of noise is to consider word probability of the given character sequence and its overlapping strings [6]. Given character sequence $s = c_0...c_n$, the left overlapping string of s is $s_L = c_{-m}...c_0...c_n$, the $c_{-m}...c_{-1}$ is the m-character preceding sequence of s, the value of m is in the set $\{1,...,n\}$ and denote the set of k-character preceding sequence as S_L. The left overlapping score of s is then calculated as Eq. 8.

$$OV_{left}(s) = \frac{\sum\limits_{s_L \in S_L(s)} I(s_L)}{|S_L(s)|} \tag{8}$$

Here $I(\cdot)$ is the indicator function defined as Eq. 9

$$I(s_L) = \begin{cases} 0 & P_{word}(s_L) \leq P_{word}(s) \\ 1 & P_{word}(s_L) > P_{word}(s) \end{cases} \tag{9}$$

The right overlapping score of s, $OV_{right}(s)$, can be calculated similarly. The overall overlapping score is defined as Eq. 10.

$$OV(s) = max\{OV_{left}(s), OV_{right}(s)\} \tag{10}$$

Character sequences with OV value larger than certain threshold are eliminated from the set of valid word candidates. A dictionary will serve as the knowledge base, any word in the dictionary will be consider as its $P_{word}(\cdot)$ is ∞. Then take the "国媒体" (country media) case as an example again, "美国(American)", "中国(China)", "美国媒体(American media)", etc. are elements in S_L(国媒体) such that the left overlapping score of "国媒体" is high. That is because P_{word}(美国) $> P_{word}$(国媒体), P_{word}(中国) $> P_{word}$(国媒体), P_{word}(美国媒体) $> P_{word}$(国媒体) and we can find many such cases using character sequences in S_L(国媒体) so the value OV(国媒体) is large which helps to identify "国媒体" as a wrong segmentation.

Overall Procedure of New Word Detection. Generally speaking, most of the Chinese words contain 2–4 characters, so we just consider character sequences whose length between two and four in the tweet corpus. We first select all the frequent character sequences, then calculate word probability of these character sequences bases on their SCP and BE score. Noises are filtered by the OV score of the character sequences afterward. Figure 1 shows an overview of the process of new word detection.

And the pseudo code of the procedure of new word detection is listed in Algorithm 1. Recall $Set_{tweet}(t)$ is the tweet corpus at time t, $Set_{tweet}(t_0)$ is the tweet corpus at time t_0 and let $Thres_{freq}$, $Thres_{wordprob}$ and $Thres_{OV}$ denotes the thresholds of character frequency, word probability score and overlapping score respectively.

5 New Word Tagging

The proposed idea of new word interpretation is automatic annotating a new word by tagging it with known words. Tagging is being extensively used in images (photos on facebook) [10] or articles annotation [11]. The objective of new word tagging is to shed light on the meaning of the word and facilitate users' better understanding. The objective of new word tagging is to shed light on the meaning of the word and facilitate users' better understanding. Our idea is to find out a list of known words that are most relevant to the given new word. Words in the following categories are potential tag words:

- Words that are highly relevant to w_{new}, i.e. its attributes, category and related named entities.
- Words that are semantically similar to w_{new}, i.e. synonyms.

The first category of words is important for tagging new words related to certain social events. It may include people, organizations or microblog user's

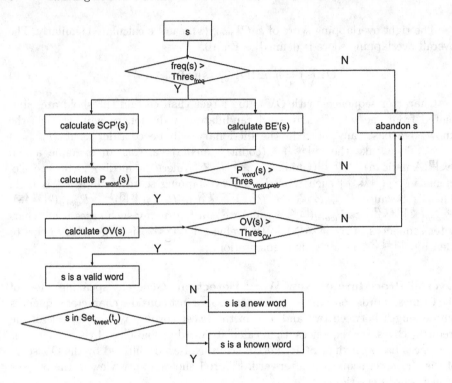

Fig. 1. Process of new word detection

comments which relate to the event. Normally, these words frequent co-occur with w_{new}. On the other hand, the second category, w_{new}'s synonyms, may not co-occur with w_{new} since the users often choose either w_{new} or its synonyms, but not both of them in the same tweet. For instance, "火星哥" (Mars brother) is a nickname of "华晨宇" (Hua Chenyu, the name of a singer) to indicate his abnormal behavior. These two terms are related but do not co-occur frequently in tweets because they can be a replacement for each other. Thus, we further quantify the similarity of two words by modeling the similarity of their corresponding contexts. The context of a word w is the surrounding text of w, roughly speaking, two words that share similar contexts are highly relevant.

Given a new word w_{new}, we select the known words which are similar to w_{new} as its tag words. The similarity evaluation often falls into two subcategories, one is distance based and the other is angle based. In this paper, we propose two approaches, namely Context Distance and Context Cosine Similarity, to measure the similarity of the new words and known words from these two aspects. Given a new word, Context Distance finds tags of the new word by the distance between the known words and the new word while Context Cosine Similarity selects tag words based on the angle between the vectors of the known words and the given new word. The result shows Context Cosine Similarity can pick tag words more precisely compares to that of Context Distance.

Algorithm 1. New Word Detection

1: **for all** $s \in Set_{tweet}(t), 2 \leq |s| \leq 4$ **do**
2:     **if** $freq(s) \geq Thres_{freq}$ **then**
3:         Calculate $SCP(s)$ using Eq. 1
4:         Get the normalized $SCP(s)$, $SCP'(s)$, using Eq. 7
5:         Calculate $BE(s)$ using Eq. 4
6:         Get the normalized $BE(s)$, $BE'(s)$, using Eq. 6
7:         Get the word probability score $P_{word}(s)$ using Eq. 5
8:         **if** $P_{word}(s) \geq Thres_{wordprob}$ **then**
9:             Add s to word candidate set Set_{cand}
10:         **end if**
11:     **end if**
12: **end for**
13: **for all** $s \in Set_{cand}$ **do**
14:     Calculate $OV(s)$ using Eq. 10
15:     **if** $OV(s) < Thres_{OV}$ **then**
16:         **if** $s \notin Set_{tweet}(t_0)$ **then**
17:             Add s to Set_{new}
18:         **else**
19:             Add s to Set_{known}
20:         **end if**
21:     **end if**
22: **end for**

5.1   Context Distance

From our study, the surrounding text of a word may shed light on its meaning. We could simply model the context of w_{known} as the set of words co-occurring with w_{new} in $Set_{tweet}(t)$. Let $doc(w_1, w_2)$ denotes the pseudo document made by concatenation of all tweets containing w_1 while w_1, w_2 are excluded from the document, we can get $doc(w_{new}, w_{known})$ and $doc(w_{known}, w_{new})$ using all the tweets containing w_{new} and w_{known} respectively. For example, the new word w_{new} is represented by a vector $v(w_{new}) = v_1, v_2, ..., v_k$, the ith element $v_i(w_{new})$ in $v(w_{new})$ indicates the relationship between w_{new} and the ith known word w_i. The value of $v_i(w_{new})$ is defined by Eq. 11. The length of $v(w_{new})$, denoted as k, equals to the size of the Set_{known}.

$$v_i(w_{new}) = \frac{tf(w_i, doc(w_{new}, w_i)) \times idf(w_i, D)}{|doc(w_{new}) \in D|} \tag{11}$$

The numerator is the term frequency - inverse document frequency (TF-IDF) weight of w_i which is a numerical statistic that is intended to reflect how important w_i is to $doc(w_{new}, w_i)$ in the tweet corpus D. Here we use $Set_{tweet}(t_0)$ as our tweet corpus and we use the occurrence of w_i as its term frequency (Eq. 12) and the proportion of tweets containing w_i in $doc(w_{new}, w_i)$ (before w_i is excluded from the document) is the inverse document frequency of w_i (Eq. 13).

The denominator is the number of tweets containing w_{new} among all the tweets.

$$tf(w_i, doc(w_{new}, w_{known})) = \frac{freq(w_i, doc(w_{new}, w_i))}{\sum\limits_{w \in doc(w_{new}, w_i)} freq(w, doc(w_{new}, w_i))} \qquad (12)$$

$$idf(w_i, D) = \log \frac{|D|}{|doc(w_i) \in D|} \qquad (13)$$

Where $doc(w_i)$ is all the tweets containing w_i.

Similarly, we can vectoring a known word w_{known} with $doc(w_{known}, w_{new})$. It is worthy noting that w_{new} and w_{known} are excluded from $doc(w_{new}, w_{known})$ and $doc(w_{known}, w_{new})$ because we assume if two words are semantically similar, their context should be similar even they co-occur with low frequency.

After using $v(w_{new})$ and $v(w_{known})$ to represent the new word and the known word as two vectors in high dimensional space, we can evaluate the similarity of w_{new} and w_{known} using their Context Distance. The proposed Context Distance is calculated by Euclidean Distance (Eq. 14) of the two words. The w_{new}, w_{known} are similar if the $dist(w_{new}, w_{known})$ is small. k equals to the size of the known word set.

$$dist(w_{new}, w_{known}) = \sqrt{\sum_{i=1}^{k} (v_i(w_{new}) - v_i(w_{known}))^2}. \qquad (14)$$

The overall procedure of calculating Context Distance is listed in Algorithm 2.

5.2   Context Cosine Similarity

Different from Context Distance which evaluating the similarity of the new word and known the word by data point distance, Context Cosine Similarity define their similarity by the angle between the new word and known word vectors. The new word and the known word are similar if the angle between them is small.

Let $v''(w_{new})$ denotes the new word vector of Context Cosine Similarity, the ith element $v_i''(w_{new})$ in $v(w_{new})$ is defined by Eq. 15.

$$v_i(w_{new}) = tf(w_i, doc(w_{new}, w_i)) \qquad (15)$$

The vector of known word w_{known}, $v''(w_{known})$, can be generated similarly. The angle between the two words is indicated by the cosine similarity of $v''(w_{new})$ and $v''(w_{known})$ (Eq. 16).

$$sim(w_{new}, w_{known}) = \frac{v''(w_{new}) \cdot v''(w_{known})}{|v''(w_{new})||v''(w_{known})|}$$

$$= \frac{\sum\limits_{i=1}^{k} v_i''(w_{new}) \times v_i''(w_{known})}{\sqrt{\sum\limits_{i=1}^{k} v_i''(w_{new})^2} \times \sqrt{\sum\limits_{i=1}^{k} v_i''(w_{known})^2}} \qquad (16)$$

$sim(w_{new}, w_{known})$ will be a value between 0 and 1, the higher the value is, the new word and the known word are more relevant.

Algorithm 2. Calculate Context Distance

1: Given a new word w_{new} and a known word w_{known}
2: **for all** $T \in Set_{tweet}(t)$ **do**
3:    **if** T contains w_{new} **then**
4:        Add T to $doc(w_{new})$
5:    **end if**
6:    **if** T contains w_{known} **then**
7:        Add T to $doc(w_{known})$
8:    **end if**
9: **end for**
10: Get $doc(w_{new}, w_{known})$ by excluding all w_{new} and w_{known} from $doc(w_{new})$
11: Get $doc(w_{known}, w_{new})$ by excluding all w_{known} and w_{new} from $doc(w_{known})$
12: **for all** $w_i \in Set_{known}$ **do**
13:    Count $freq(w_i, doc(w_{new}, w_i))$
14:    Count $freq(w_i, doc(w_i, w_{new}))$
15:    **for all** $T \in Set_{tweet}(t)$ **do**
16:        **if** T contains w_i **then**
17:            Put T into $doc(w_i)$
18:        **end if**
19:    **end for**
20:    Calculate $idf(w_i, Set_{tweet}(t))$ using Eq. 13)
21:    Calculate $v_i(w_{new}), v_i(w_{known})$ using Eq. 11
22: **end for**
23: Calculate $dist(w_{new}, w_{known})$ using Eq. 14

5.3 Choose Tag Words

Since the Context Distance and Context Cosine Similarity is relevant to the number of known words occur in $doc(w_{new}, w_{known})$, the average value and the standard variation is different for distinct new words. In other words, we cannot set a unified threshold for all the tag words of different new words directly. In this case, we perform a further max-min normalization on the top 20 relevant tag word of a specific new word. The max-min normalization formula is as Eq. 17.

$$dist'(i) = \frac{dist(i) - min_{dist}}{max_{dist} - min_{dist}} \tag{17}$$

After the max-min normalization, we can set a unified threshold $Thres_{CD}$ for all the new words in tag words selecting. The process of choosing tag words bases on Context Cosine Similarity is similar, but sorting the values in Set_{sim} by value descending then choosing the largest 20 values and w_{known} is considered as a tag word when the further normalized Context Cosine Similarity $sim'(w_{known})$ is larger than a threshold $Thres_{CCS}$.

The process of selecting tag words bases on Context Distance is listed as Algorithm 3.

The process of selecting tag words bases on Context Cosine Similarity is similar, but it is worth noticing we assuming w_{new} and w_{known} are relevant when the Context Distance $dist(w_{new}, w_{known})$ is small while assuming they are

---

Algorithm 3. Choose Tag Words Base on Context Distance

1: Given a new word w_{new}
2: **for all** $w_{known} \in Set_{known}$ **do**
3:      Using Eq. 14 to get $dist(w_{new}, w_{known})$
4:      Add $dist(w_{new}, w_{known})$ to Set_{dist}
5: **end for**
6: Sort Set_{dist} by its value ascending and choose the top 20 values as Set'_{dist}
7: Select the max value in Set'_{dist} as max_{dist}
8: Select the minimal value in Set'_{dist} as min_{dist}
9: **for all** $dist \in Set'_{dist}$ **do**
10:      Using Eq. 17 to normalize the distance, $dist'(w_{new}, w_{known})$ is in range [0,1]
11:      **if** $dist'(w_{new}, w_{known}) \leq Thres_{CD}$ **then**
12:          w_{known} is selected as a tag word of w_{new}
13:      **end if**
14: **end for**

---

relevant when the Context Cosine Similarity $sim(w_{new}, w_{known})$ is large, so we should sort the elements in the CCS value set Set_{sim} by value descending rather than ascending.

6   Experiment

6.1   Dataset Setting

In this experiment, we aim at detecting newly emerged words on a daily basis. Regarding to the definition of new words, for the target day t, $Set_{tweet}(t)$ is the set of tweets published on that day. Tweets published in seven consecutive days, from July 31st, 2013 to Aug 6th, 2013 are used as our input. Meanwhile, we use the tweets of May 2013 as the known word set $Set_{tweet}(t_0), t_0 < t$ which serves as knowledge base. Hash tags, spam tweets and tweets only contains non-Chinese characters are filtered in the dataset. Table 2 shows the details of our dataset. And we store any character sequence with length between two and four in $Set_{tweet}(t_0)$ to serve as the known word set $Set_{word}(t_0)$ to ensure new words detected from $Set_{tweet}(t)$ has never appeared in $Set_{tweet}(t_0)$.

We perform cleaning on dataset used as $Set_{tweet}(t)$, where hash tags, spam tweets, tweets only contains non-Chinese characters are rejected. We store any character sequence with length between two and four in $Set_{tweet}(t_0)$ to serve as the known word set $Set_{word}(t_0)$ to ensure new words detected from $Set_{tweet}(t)$ has never appeared in $Set_{tweet}(t_0)$. Table 2 shows the details of our dataset. From Table 2, we can see that there are over 20 million tweets of $Set_{tweet}(t_0)$, around 50 times larger than the size of $Set_{tweet}(t)$, such that we assume $Set_{tweet}(t_0)$ is sufficient to server as our knowledge base. Moreover, even there is any word do not exist in $Set_{tweet}(t_0)$ but frequently appear in $Set_{tweet}(t)$, we assume the word becomes hot again for a new reason, this type of words also deserves our attention and needs to be annotated.

Table 2. List of dataset

Dataset	# of tweets	After cleaning
July 31	715, 680	443,734
Aug 1	824, 282	515,837
Aug 2	829, 224	516,152
Aug 3	793, 324	397,291
Aug 4	800, 816	392,945
Aug 5	688, 692	321,341
Aug 6	785, 236	399,699
May	20, 700, 001	-

6.2   New Word Detection Result

In the new word detection experiment, we use ICTCLAS and Stanford Chinese-word-segmenter [21] to serve as our baselines. The training data used by ICT-CLAS is Peking University dataset which contains 149,922 words while training data used for CRF training is Penn Chinese Treebank which contains 423,000 words. All the words appearing in the training set will not be selected as a new word. Non-Chinese character, emotion icon, punctuation, date, word containing stop words and some common words are excluded because they are not our target. And our aim is to detect new words of certain importance as well as their relevant words, it is reasonable to focus on words with relatively high frequency. In this experiment, we just evaluating the word probability of the top 5% frequent character sequence and set the threshold $Thres_{freq}$ to 15, words appearing less than $Thres_{freq}$ will be ignored. Figure 2 shows the character sequence amount of different frequency. We can see that most of the character sequences occur less than 5 times. However, from Fig. 3 we can notice that the low-frequency character sequences often not valid words. Take the character sequences occur less than 5 times as an example, only 5% of them are valid words.

Generally speaking, Chinese new words can be divided into several categories [20] (excluding new words with non-Chinese characters): name entity, dialect, glossary, novelty, abbreviation and transliterated words. The detected new words are classified according to these categories in our experiment. The precision of the detection result is defined as Eq. 18

$$Precision = \frac{\text{\# of valid new words}}{\text{\# of total new words detected}} \tag{18}$$

The threshold of overlapping score, $Thres_{OV}$, is set to 0.7, same as that in [6]. We set the threshold of the word probability, $Thres_{wordprob}$, to 0.3 bases on the following observation: the word probability of all the character sequences mainly fall into the range [0, 1] and there are around 30% of the character sequences whose frequency larger than $Thres_{freq}$ are an invalid word.

The experiment results are listed in Table 3.

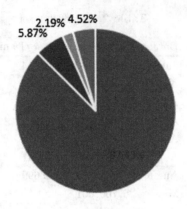

Fig. 2. Character sequence proportion of different frequency

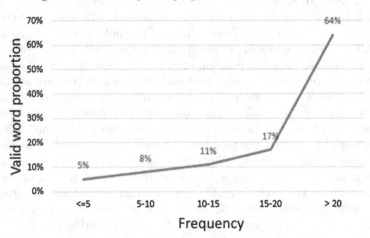

Fig. 3. Valid word proportion of different frequency

The results show that our method has the highest precision in detecting new words in Chinese Twitter among the three methods. Stanford Chinese-word-segmenter wins in recall. However, a large number of noise is also included in Stanford Chinese-word-segmenter's result which lowers the precision tremendously. The reason is that it uses a supervised machine learning method, for which the shortage of appropriate tagged training dataset for Chinese tweet is a fatal problem. ICTALS has an acceptable precision, but it often over segment the words which makes it fails to detect some compound words such as *"Yu'E Bao"* and *"Wechat Wo card"*. For example, for the sentence "余额宝将让中国银行倒闭" (*Yu'E Bao* will make Chinese banks bankrupt), "余额宝" (*Yu'E Bao*) is a totally new product while "余额" is a

Table 3. New word detection result

Category	Our method	ICTCLAS	Chinese-word-segmenter
Name Entity	50	36	60
Glossary	2	1	6
Novelty	19	2	36
Abbreviation of hot topic	22	1	8
Transliterated words	0	0	5
Noise	4	5	139
Valid new words	93	40	**115**
Precision	**95.9%**	88.9%	45.2%

known word which probably exists in the previous corpus, the segmentation result constructed by ICTCLAS is 余额/宝/ 将/让/ 中国/ 银行/倒闭 which separates "余额/宝/ 将/让/ 中国/ 银行/倒闭" (Yu'E) and "宝" (Bao) as two different words. To summarize, our method is more effective than the two baselines in term of recall and precision as a whole.

6.3 Tagging Result Bases on Context Distance

Among the 93 detected new words, some of them are recorded in Baidu Entry now. We randomly picked 20 recorded words from different categories to evaluate our tagging result. The precision of tagging result about a new word (w_{new}) is defined as:

$$Precision_{tag}(w_{new}) = \frac{\text{\# of tag words hits in } w_{new}\text{'s Baidu Entry}}{\text{\# of } w_{new}\text{'s tag words}} \quad (19)$$

Known words whose normalized Context Distance lower than a threshold $Thres_{CD}$ are selected as the tag of the new word. Words such as 加油 (work hard) and 执行 (operate) are excluded in tag words manually since they are either only popular in Sina Microblog or do not have much meaning on its own. We have tried different $Thres_{CD}$ and compare the tagging accuracy and the number of tag words is selected. The result is in Table 4.

Table 4. Word tagging result of different $Thres_{CD}$

Threshold	Average tag count	Average precision
0.25	2.2	0.65
0.5	4.4	0.55
0.75	8.3	0.52
1	20	0.44

From Table 4 we can see that the number of selected tag words decreases while the tagging precision increase when $Thres_{CD}$ decline. This is in consistent with tag words which have smaller Context Distance with the new word is more likely be the right tag. Generally, we seek for high tagging accuracy while need enough number of tags to have a detailed interpretation of the given new word. We can achieve the highest precision when we set $Thres_{CD} = 0.25$, however, only 2.2 known words are selected as tag words of the given new word in this case which make the interpretation indistinctly.

According to Table 4, we set $Thres_{CD} = 0.75$ to get a good balance between the tagging accuracy and the number of tag words. Some tagging result are listed as below.

- **Yu'E Bao** (Novelty. Yu'E Bao is a money-market fund promoted by Alipay)
  Tags: Suning Online Market (An online shop which can using Yu'E Bao to pay), fund, money management
- **Wechat Wo card** (Novelty. A SIM card released by Tencent and China Unicom. People using Wo card can have some special rights in Wechat)
  Tags: special right, China Unicom, Wechat, Tencent
- **Liu Yexing** (Name Entity. A guy from Tsinghua University become famous by attending reality show "Who's still standing" and burst their question bank)
  Tags: idol, China, sound, succeed, summer, young, end, perfect
- **Wu Mulan** (Name Entity. A player in the TV program "The voice of China". An infertility patients claimed she get pregnant after listen to Wu's song since the music made she relax)
  Tags: voice, China, song, enjoy, music, pregnant, view, sprit, child, support, talk, strong, sing
- **Ergosterol** (Glossary. Ergosterol is a sterol found in cell membranes of fungi and protozoa, serving many of the same functions that cholesterol serves in animal cells)
  Tags: eat frequently, growth, mushroom, virus, immunity
- **Burst Bar event** (Abbreviation of hot topic. Pan Mengyin, a fan of Korean star G-Dragon, spread some inappropriate remarks about football stars on Internet which makes fans of the football stars get angry and attacked G-Dragon's Baidu Bar)
  Tags: G-Dragon, hope, friend, whole life, Pan Mengyin, Internet, strong, birthday, strength

Moreover, we have further compare the number of tags and the tagging precision of different word categories. The result is in Table 5.

6.4 Tagging Result Bases on Context Cosine Similarity

Known words whose normalized Context Cosine Similarity higher than a threshold $Thres_{CCS}$ are selected as the tag of the new word. The tagging accuracy and the number of tag words of different $Thres_{CCS}$ is shown in Table 6.

Table 5. Word tagging result of different categories

Category	New word count	Average tag count	Average precision
Name entity	9	9.3	0.45
Glossary	1	5.0	0.00
Novelty	4	7.5	0.62
Abbreviation of hot topic	6	7.8	0.66

Table 6. Word tagging result of different thresholds

Threshold	Average tag count	Average precision
0	19.6	0.56
0.25	9.1	0.71
0.5	5.5	0.79
0.75	3	0.825

Table 6 shows the number of selected tag words decreases while the tagging precision increase when the threshold of Context Cosine Similarity arise. This indicates the tag words which have higher context cosine similarity with the new word is more likely be the right tag of the new word. In our experiment, we set the threshold $Thres_{CCS} = 0.5$ such that we can get enough tag words while achieving a relatively high precision. Some tagging result are listed as below.

- **Yu'E Bao** (Novelty. Yu'E Bao is a money-market fund promoted by Alipay)
  Tags: currency, Internet, finance, fund, money management, supply-chain
- **Wechat Wo card** (Novelty. A SIM card released by Tencent and China Unicom. People using Wo card can have some special rights in Wechat)
  Tags: special right, China Unicom, Tencent, network traffic
- **Liu Yexing** (Name Entity. A guy from Tsinghua University become famous by attending reality show "Who's still standing" and burst their question bank)
  Tags: Zhang Xuejian (Liu Yexing's adversary), Who's still standing, Tsinghua University, Peking University, question bank, answer questions
- **Wu Mulan** (Name Entity. A player in the TV program "The voice of China". An infertility patients claimed she get pregnant after listen to Wu's song since the music made she relax)
  Tags: Mushroom Brothers, Liu Zichen, Yu Junyi, Ta Siken, Ni peng, supervisor (The first five tags are other players in the same show, the last tag is a role in the show)
- **Ergosterol** (Glossary. Ergosterol is a sterol found in cell membranes of fungi and protozoa, serving many of the same functions that cholesterol serves in animal cells)
  Tags: protein, amount, immunity, vegetables, growth

- **Burst Bar event** (Abbreviation of hot topic. Pan Mengyin, a fan of Korean star G-Dragon, spread some inappropriate remarks about football stars on Internet which makes fans of the football stars get angry and attacked G-Dragon's Baidu Bar)
  Tags: Pan Mengyin, G-Dragon, Korean star, stupid

We also further compared the number of tags and tagging precisions of different word categories in Table 7).

Table 7. Word tagging result of different categories

Category	# of new words	Average # of tags	Average precision
Name entity	9	6.11	0.80
Glossary	1	6.00	0.00
Novelty	4	3.00	0.96
Abbreviation of hot topic	6	6.17	0.79

6.5   Tagging Result Discussion

Comparing the tagging result bases on Context Distance and that of Context Cosine Similarity, we found Context Cosine Similarity performs better among the two approaches. By comparing Tables 4 and 6, we can notice that for a similar number of selected tag words, Context Cosine Similarity often gets a higher precision. For example, in Table 4, when $Thres_{CD} = 1$, there are 20 tag words selected for each new word on average, the precision of the tag word selection is 0.44. That compares to $Thres_{CCS} = 0$ in Table 6, there are 19.1 tag words selected for each new word on average, the precision of the tag word selection achieves 0.56. This phenomena might because in Context Distance calculation, although tf-idf has been utilized to reduce the impact of the imbalance of word frequency, some common words are still selected as a tag word of a new word even they are not highly relevant (e.g. "friend" and "birthday" are selected as the tag words of "*Pan Mengyin*"). Context Cosine Similarity overpass the word frequency imbalance problem by using data distribution to define the similarity value.

By analyzing the word tagging results of different categories (Tables 5 and 7), we can see an interesting thing that comparing to name entity and abbreviation of hot topic, novelty have fewer number of tag words while achieves much higher precision. Both of Context Distance and Context Cosine Similarity failed in tagging the glossary *Ergosterol* precisely because a lot of tweets talking *Ergosterol* are a kind of advertisement. Moreover, even some related words are selected as a tag word of *Ergosterol*, the online encyclopedia Baidu Entry using more technical terms to explain *ergosterol*. For example, mushroom is a tag word of *Ergosterol* bases on Context Distance and mushroom contains a lot of Ergosterol so these two words are relevant, but in Baidu Entry, it used the term *Fungus* instead of mushroom.

7   Conclusion and Future Work

In this paper, we consider the problem of detecting and interpreting new words in Chinese Twitter. We proposed an unsupervised new word detection framework which take several statistical features to derive a word probability score that can measure word-forming likelihood of a character sequence. The proposed framework detect new words based on statistics information such as sequence frequency, the specific wording do not serve as features in this case, so it could be easily applied to other Kanji based languages (e.g. Japanese and Korean).

Then, we used automatic tagging in new word interpretation. We derive a similarity measure between new word and its candidate tag word based on similarity of their corresponding contexts. Experiments on real datasets show the effectiveness of our approach. However, in this work, some thresholds, such as $freq(\cdot)$ and $Pr_{word}(s)$, are set by experiments and observation. In real practise, we can have a more systematic and statistical way to set some appropriate thresholds. For example, for the frequency, we can compute the mean and the standard deviation of the identified words, then set a threshold based on the mean and the standard deviation. In the future, we will try to explore an automatic way to define the parameters used in this framework and apply the language model in our research to get more accurate results.

References

1. Peng, F., Feng, F., McCallum, A.: Chinese segmentation and new word detection using conditional random fields. In: Proceedings of the 20th International Conference on Computational Linguistics, p. 562. Association for Computational Linguistics (2004)
2. Finin, T., et al.: Annotating named entities in Twitter data with crowdsourcing. In: Proceedings of the NAACL HLT 2010 Workshop on Creating Speech and Language Data with Amazon's Mechanical Turk. Association for Computational Linguistics (2010)
3. Ritter, A., Clark, S., Etzioni, O.: Named entity recognition in tweets: an experimental study. In: Proceedings of the Conference on Empirical Methods in Natural Language Processing. Association for Computational Linguistics (2011)
4. Gattani, A., et al.: Entity extraction, linking, classification, and tagging for social media: a wikipedia-based approach. Proc. VLDB Endow. **6**(11), 1126–1137 (2013)
5. Sun, X., Wang, H., Li, W.: Fast online training with frequency-adaptive learning rates for Chinese word segmentation and new word detection. In: Proceedings of the 50th Annual Meeting of the Association for Computational Linguistics: Long Papers, vol. 1. Association for Computational Linguistics (2012)
6. Ye, Y., Qingyao, W., Li, Y., Chow, K.P., Hui, L.C.K., Kwong, L.C.: Unknown Chinese word extraction based on variety of overlapping strings. Inf. Process. Manag. **49**(2), 497–512 (2013)
7. Zhao, H., Kit, C.: Exploiting unlabeled text with different unsupervised segmentation criteria for Chinese word segmentation. Res. Comput. Sci. **33**, 93–104 (2008)
8. Nadeau, D., Sekine, S.: A survey of named entity recognition and classification. Lingvisticae Investigationes **30**(1), 3–26 (2007)

9. Zhao, H., Kit, C.: Unsupervised segmentation helps supervised learning of character tagging for word segmentation and named entity recognition. In: IJCNLP, pp. 106–111 (2008)

10. Zhou, N., et al.: A hybrid probabilistic model for unified collaborative and content-based image tagging. IEEE Trans. Pattern Anal. Mach. Intell. **33**(7), 1281–1294 (2011)

11. Kim, H.-N., et al.: Collaborative filtering based on collaborative tagging for enhancing the quality of recommendation. Electron. Commer. Res. Appl. **9**(1), 73–83 (2010)

12. Luo, S., Sun, M.: Two-character Chinese word extraction based on hybrid of internal and contextual measures. In: Proceedings of the Second SIGHAN Workshop on Chinese Language Processing, vol. 17. Association for Computational Linguistics (2003)

13. Jin, Z., Tanaka-Ishii, K.: Unsupervised segmentation of Chinese text by use of branching entropy. In: Proceedings of the COLING/ACL on Main Conference Poster Sessions, Association for Computational Linguistics (2006)

14. Wang, L., et al.: CRFs-based Chinese word segmentation for micro-blog with small-scale data. In: Proceedings of the Second CIPS-SIGHAN Joint Conference on Chinese Language (2012)

15. Zhang, K., Sun, M., Zhou, C.: Word segmentation on Chinese mirco-blog data with a linear-time incremental model. In: Proceedings of the Second CIPS-SIGHAN Joint Conference on Chinese Language Processing, Tianjin (2012)

16. Zhang, H.-P., et al.: HHMM-based Chinese lexical analyzer ICTCLAS. In: Proceedings of the Second SIGHAN Workshop on Chinese Language Processing, vol. 17. Association for Computational Linguistics (2003)

17. Dumais, S.T.: Latent semantic analysis. Annu. Rev. Inf. Sci. Technol. **38**(1), 188–230 (2004)

18. Aksoy, S., Haralick, R.M.: Feature normalization and likelihood-based similarity measures for image retrieval. Pattern Recognit. Lett. **22**(5), 563–582 (2001)

19. Kityz, C., Wilksz, Y.: Unsupervised learning of word boundary with description length gain. In: Proceedings of the CoNLL99 ACL Workshop. Association for Computational Linguistics, Bergen (1999)

20. Gang, Z., et al.: Chinese new words detection in internet. Chin. Inf. Technol. **18**(6), 1–9 (2004)

21. Tseng, H., et al.: A conditional random field word segmenter for sighan bakeoff 2005. In: Proceedings of the Fourth SIGHAN Workshop on Chinese Language Processing, Jeju Island, vol. 171 (2005)

22. Zheng, X., Chen, H., Xu, T.: Deep learning for Chinese word segmentation and POS tagging. In: EMNLP (2013)

23. Wallach, H.M.: Conditional random fields: an introduction (2004)

On-Demand Snapshot Maintenance in Data Warehouses Using Incremental ETL Pipeline

Weiping Qu[✉] and Stefan Dessloch

Heterogeneous Information Systems Group,
University of Kaiserslautern, Kaiserslautern, Germany
{qu,dessloch}@informatik.uni-kl.de

Abstract. Multi-version concurrency control method has nowadays been widely used in data warehouses to provide OLAP queries and ETL maintenance flows with concurrent access. A snapshot is taken on existing warehouse tables to answer a certain query independently of concurrent updates. In this work, we extend the snapshot in the data warehouse with the deltas which reside at the source side of ETL flows. Before answering a query which accesses the warehouse tables, relevant tables are first refreshed with the exact source deltas which are captured until this query arrives and haven't been synchronized with the tables yet (called on-demand maintenance). Snapshot maintenance is done by an incremental recomputation pipeline which is flushed by a set of consecutive, non-overlapping delta batches in delta streams which are split according to a sequence of incoming queries. A workload scheduler is thereby used to achieve a serializable schedule of concurrent maintenance jobs and OLAP queries. Performance has been examined by using read-/update-heavy workloads.

1 Introduction

Nowadays companies are emphasizing the importance of data freshness of analytical results. One promising solution is executing both OLTP and OLAP workloads in a 1-tier *one-size-fits-all* database such as Hyper [8], where operational data and historical data reside in the same system. Another appealing approach used in a common 2-tier or 3-tier configuration is near real-time ETL [1] by which data changes from transactions in OLTP systems are extracted, transformed and loaded into the target warehouse in a small time window (five to fifteen minutes) rather than during off-peak hours. Deltas are captured by using Change-Data-Capture (CDC) methods (e.g. log-sniffing or timestamp [10]) and propagated using incremental recomputation techniques in micro-batches.

Data maintenance flows run concurrently with OLAP queries in near real-time ETL, in which an intermediate Data Processing Area (DPA, as counterpart of the data staging area in traditional ETL) is used to alleviate the possible overload of the sources and the warehouse. It is desirable for the DPA to relieve *traffic jams* at a high update-arrival rate and meanwhile at a very high query rate alleviate the burden of locking due to concurrent read/write accesses to

© Springer-Verlag GmbH Germany 2017
A. Hameurlain et al. (Eds.): TLDKS XXXII, LNCS 10420, pp. 91–112, 2017.
DOI: 10.1007/978-3-662-55608-5_5

shared data partitions. Alternatively, many data warehouses deploy the Multi-Version Concurrency Control (MVCC) mechanism to solve concurrency issues. If serializable snapshot isolation is selected, a snapshot is taken at the beginning of a query execution and used during the entire query lifetime without interventions incurred by concurrent updates. For time-critical decision making, however, the snapshot taken at the warehouse side is generally stale since at the same moment, there could be deltas that haven't captured yet from the source OLTP systems or being processed by an ETL tool. In order to achieve a more refreshed snapshot, it is needed to first synchronize the source deltas with the relevant tables before taking the snapshot. Hence, a synchronization delay cannot be avoided which is incurred by an ETL flow execution.

The scope of our work is depicted in Fig. 1. We assume that a CDC process runs continuously and always pulls up-to-date changes without those maintenance anomalies addressed in [5].

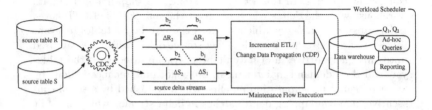

Fig. 1. Consistency scope in warehouse model

Correct and complete sets of delta tuples (Δ: insertions, updates and deletions on source tables R and S) are continuously pushed into so-called *source delta streams* in DPA. An event of a query arrival at the warehouse side triggers the system to group the current delta tuples in every source delta stream as a *delta batch* and to construct a *maintenance job* which takes the delta batches as input and perform one run of maintenance flow execution using incremental ETL techniques. The final delta batch which is produced by this maintenance flow execution is used to refresh the target warehouse tables and then a snapshot is taken for answering this query. For example, as shown in Fig. 1, the arrival of Q_1 leads to the construction of a maintenance job m_1. The input for m_1 are two delta batches b_1 with the delta tuples as ΔR_1 and ΔS_1 that are derived from the source transactions committed before the arrival time of Q_1. The query execution of Q_1 is initially suspended and later resumed when relevant tables are refreshed by the output of m_1. We call this *on-demand maintenance policy*.

With a sequence of incoming queries, a list of chained maintenance jobs are created for ETL flow to process. For efficiency and consistency, several challenges exist and are listed as follows. Firstly, sequential execution of ETL flow instances can lead to high synchronization delay at a high query rate. Parallelism needs to be exploited at certain level of the flow execution to improve performance. Furthermore, general ETL flows could contain operations or complex user-defined procedures which read and write shared resources. While running separate ETL flow/operation instances simultaneously for different maintenance jobs, inconsistency may occur due to uncontrolled access to shared resources. Finally, in our work, a warehouse snapshot S_i is considered as *consistent* for an incoming query Q_i if S_i is contiguously updated by final delta batches from preceding maintenance jobs (m_1-m_i) before the submission time of Q_i and is not interfered by fast finished succeeding jobs (e.g. m_{i+1}, which leads to non-repeatable read/phantom read anomalies). While timestamps used to extend both delta tuples and target data partitions could be a possible solution to ensure query consistency, this will result in high storage and processing overheads. A more promising alternative is to introduce a mechanism to schedule the update sequence and OLAP queries.

In this work, we address the real-time snapshot maintenance problem in MVCC-supported data warehouse systems using near real-time ETL techniques. The objective of this work is to achieve high throughput at a high query rate and meanwhile ensure the serializability property among concurrent maintenance flow executions in ETL tools and OLAP queries in warehouses. The contributions of this work are as follows:

- We introduce our on-demand maintenance policy for snapshot maintenance in data warehouses according to a computational model.
- Based on the infrastructure introduced for near real-time ETL [1], we proposed for an incremental ETL pipeline as a runtime implementation of the logical computational model using an open-source ETL tool called Pentaho Data Integration (Kettle) (shortly Kettle) [10]. The incremental ETL (job) pipeline can process a list of chained maintenance jobs simultaneously for high query throughput.
- We define the consistency notion in our real-time ETL model based on which a workload scheduler is proposed for a serializable schedule of concurrent maintenance flows and queries that avoids using timestamp-based approach. An internal queue is used to ensure consistency with correct execution sequence.
- Furthermore, we introduce *consistency zones* in our incremental ETL pipeline to avoid potential consistency anomalies, using incremental join and slowly changing dimension maintenance as examples.
- The experimental results show that our approach achieves nearly similar performance as in near real-time ETL while the query consistency is still guaranteed.

This paper is an extended version of our previous work [16] and is structured as follows. We start by introducing terminology in our work and then describe the computational model for our on-demand maintenance policy in Sect. 2. The incremental ETL pipeline is proposed in Sect. 3 as a runtime implementation of the computational model, which addresses the performance challenge. In Sect. 4, we explain the consistency model used in our work, based on which a workload scheduler is introduced in Sect. 5 to achieve the serializability property. In Sect. 6, we address potential consistency anomalies in incremental ETL pipeline and describe the consistency zones as the solutions. We validate our approach with read-/update-heavy workloads and the experimental results are discussed in Sect. 7.

2   The Computational Model

In this section, we describe the computational model for our on-demand maintenance policy. In our work, we use a dataflow system to propagate source deltas
to the data warehouse and the ETL transformation programs are interpreted as
dataflow graphs. As shown in Fig. 2, a dataflow graph is a directed acyclic graph
$G(V, E)$, in which nodes $v \in V$ represent ETL transformation operators or user-
defined procedures (in triangle form), and edges $e \in E$ are *delta streams* used to
transfer deltas from provider operators to consumer operators. A delta stream is
an ordered, unbounded collection of delta tuples (Δ: insertions (I), deletions (D)
and updates (U)) and it can be implemented as an in-memory queue, a database
table or a file. There are two types of delta streams: *source delta streams* (e.g.
streams for ΔR and ΔS) and *interior delta streams*. The source delta streams
buffer source delta tuples that are captured by an independent CDC process and
maintained in commit timestamp order in terms of source-local transactions. The
interior delta stream stores the output deltas that are processed by the provider
operator and at the same time, transfers them to the consumer operator. Hence,
the same delta stream can be either the input or the output delta stream for
two different, consecutive operators.

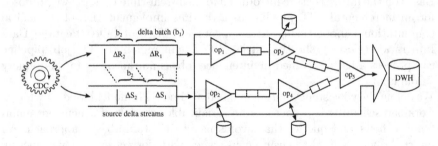

Fig. 2. Dataflow graph

Moreover, an event of a query arrival at timestamp t_i groups all source deltas
with commit-time (Δ) $< t_i$ in each source delta stream into a *delta batch* b_i and
constructs a *maintenance job* m_i. Each delta batch b_i is a finite, contiguous subsequence of a delta stream and each tuple in b_i contains not only general information for incremental processing (e.g. change flag (I, D, U), change sequence
number), but also the id of the maintenance job m_i. All the tuples in b_i have
the same maintenance job id and should be processed together as a unit in subsequent transformation operators (e.g. op_1 and op_2). The output tuples after a
delta batch processing are also assigned the same maintenance job id and are
grouped into a new delta batch for downstream processing (e.g. op_3 and op_4).
The maintenance job m_i is an abstraction of one maintenance flow execution
where all the operators in the dataflow graph process the delta batches referring
to the same job in their owning delta streams. (In the rest of the paper, we use

the terms "delta batch" and "maintenance job" interchangeably to refer to the delta tuples used in one run of each transformation operator.)

With a sequence of incoming queries, the source delta streams are split to contiguous, non-overlapping delta batches and a list of chained maintenance jobs are created for the dataflow graph to process. To deliver warehouse tables with consistent deltas, the maintenance jobs needed to be processed in order in each operator. With continuous delta batches in the input delta stream, the operator execution is deployed in the following three types, depending on how the state of (external) resources is accessed.

– For operators that only write or install updates to external resources, the operator execution on each delta batch can be wrapped into a transaction. Multiple transaction instances could be instantiated for continuous incoming delta batches and executed simultaneously while these transactions have to commit in the same order as the sequence in which the maintenance jobs are created. Transaction execution protects the state from system failure, e.g. the external state would not be inconsistent in case a system crash occurs in the middle of one operator execution with partial updates. In Fig. 2, such operators can be op_3 or op_5 which continuously update the target warehouse tables. Having multiple concurrent transaction executions on incoming delta batches with a strict commit order is useful to increase the throughput.
– For operators or more complex user-defined procedures which could both read and write the same resources, transactions run serially for incoming delta batches. For example, op_4 calculates average stock price, which needs to read the stock prices installed by the transaction executions on the preceding delta batches.
– For operators that do not access any external state or probably read a private state which is rarely mutated by other applications, no transaction is needed for the operator execution. The drawback of running a transformation operator in one transaction is that the output deltas will only be visible to downstream operator when the transaction commits. To execute operators, e.g. filter or surrogate-key-lookup (op_2), no transactions are issued. The output delta batches of these operators are generated in a tuple-by-tuple fashion and can be immediately processed by subsequent operators, thus increasing the throughput of flow execution.
– A more complicated case is that multiple separate operators could access the same shared (external) resources. Thus, additional scheduling and coordination of operator executions are needed, which is detailed in Sect. 6.

3   Incremental ETL Pipeline

As introduced before, a sequence of query arrivals force our ETL maintenance flow to work on a list of chained maintenance jobs (called *maintenance job chain*), each of which brings relevant warehouse tables to the consistent state demanded by a specific query. We address the efficiency challenge of ETL maintenance flow execution in this section. We exploit pipeline parallelism and proposed an idea of *incremental ETL pipeline*.

In more detail, we define three status of a maintenance job: *pending, in-progress* and *finished*. When the system initially starts, a pending maintenance job is constructed and put in an empty maintenance job chain. Before any query arrives, all captured source delta tuples are tagged with the id of this job. With the event of a query arrival, the status of this pending job is changed to in-progress and all delta tuples with this job id are grouped to a delta batch as input. A new pending maintenance job is immediately constructed and appended to the end of the job chain, which is used to mark subsequent incoming source deltas with this new job id. The job ids contained in the tuples from delta batches are used to distinguish different maintenance jobs executed in the incremental ETL pipeline. The ETL pipeline is an runtime implementation of the dataflow graph where each node runs in a single, non-terminating thread (*operator thread[1]*) and each edge $e \in E$ is an in-memory pipe used to transfer data from its provider operator thread to the consumer operator thread. Each transformation operator contains a pointer which iterates through the elements in the maintenance job chain. An operator thread continuously processes tuples from incoming delta batches and only blocks if its input pipe is empty or when it points at a pending job. When the job status changes to in-progress (e.g. when a query occurs), the blocked operator thread wakes up and uses the current job id to fetch delta tuples with matching job id from its input pipe. When an operator thread finishes the current maintenance job, it re-initializes its local state (e.g. cache, local variables) and tries to fetch the next (in-progress) maintenance jobs by moving its pointer along the job chain. In this way, we construct a maintenance job pipeline where every operator thread works on its own job (even for blocking operators, e.g. sort, as well). The notion of pipelining in our case is defined at job level instead of row level. However, row-level pipelining still occurs when threads of multiple adjacent operators work on the same maintenance job.

Figure 3 illustrates a state where the ETL pipeline is flushed by four maintenance jobs (m_1–m_4). These jobs are triggered by either queries or update overload[2]. At the end of this maintenance job chain exists a pending m_6 job used to assign the id of m_6 to later captured deltas. In this example, the downstream aggregation thread has delivered target deltas of m_1 to the warehouse and blocks when it tries to work on m_2 since there is still no output from its preceding (blocking) join thread. The lookup$_2$ in the bottom join branch is still working on m_2 due to slow speed or large input size while the lookup$_1$ in the upper join branch is generating output deltas of m_3. However, the deltas with the id of m_3 in the input pipe are invisible to the join thread until it finishes

---

[1] As defined in Sect. 2, there are three deployment types of operator which decide whether the operator execution in this thread is wrapped in a transaction or not.

[2] We also introduce system-level maintenance jobs which are generated when the size of an input delta stream exceeds a certain threshold without any necessary query arrival. This partially hides maintenance overhead from query response time, thus shrinking the synchronization delay for answering a late query.

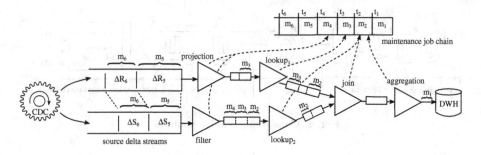

Fig. 3. Incremental ETL pipeline

m_2. Besides, a large pile-up exists in the input pipe of lookup$_2$ and more CPU cycles could be needed for it to solve transient overload. From this example, we see that our incremental ETL pipeline is able to handle continuously incoming maintenance jobs simultaneously and efficiently.

4   The Consistency Model

In this section, we introduce the notion of consistency which our work is building on. For simplicity, let us assume that an ETL flow f is given with one source table I and one target warehouse table S as sink. With an arrival of a query Q_i at point in time t_i, the maintenance job is denoted as m_i and the delta batch in the source delta stream for source table I is defined as $\Delta_{m_i} I$. After one run of maintenance flow execution on $\Delta_{m_i} I$, the final delta batch for updating the target table S is defined as follows:

$$\Delta_{m_i} S = f(\Delta_{m_i} I)$$

Given an initial state S_{old} for table S, the correct state that is demanded by the first incoming query Q_1 is derived by updating (denoted as $\uplus$) the initial state S_{old} with the final delta batch $\Delta_{m_1} S$. As defined above, $\Delta_{m_1} S$ is calculated from the source deltas $\Delta_{m_1} I$ which is captured from the source-local transactions committed before the arriving time of Q_1, i.e. t_1.

$$S_{m_1} \equiv S_{old} \uplus \Delta_{m_1} S \equiv S_{old} \uplus f(\Delta_{m_1} I)$$
$$S_{m_2} \equiv S_{m_1} \uplus \Delta_{m_2} S \equiv S_{old} \uplus \Delta_{m_1} S \uplus \Delta_{m_2} S \equiv S_{old} \uplus f(\Delta_{m_1} I) \uplus f(\Delta_{m_2} I)$$
....

$$S_{m_i} \equiv S_{m_{i-1}} \uplus \Delta_{m_i} S$$
$$\equiv S_{m_{i-2}} \uplus \Delta_{m_{i-1}} S \uplus \Delta_{m_i} S$$
...

$$\equiv S_{old} \uplus \Delta_{m_1} S \uplus \Delta_{m_2} S ... \uplus \Delta_{m_{i-1}} S \uplus \Delta_{m_i} S$$
$$\equiv S_{old} \uplus f(\Delta_{m_1} I) \uplus f(\Delta_{m_2} I) ... \uplus f(\Delta_{m_{i-1}} I) \uplus f(\Delta_{m_i} I)$$

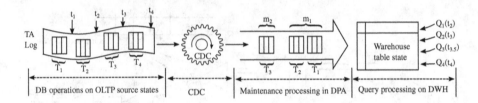

Fig. 4. Consistency model example

Therefore, we define that a snapshot of table S_{m_i} is *consistent* for the query Q_i if S_{m_i} is contiguously updated by final delta batches from preceding maintenance jobs (m_1-m_i) before the submission time of Q_i and has not received any updates from fast-finished succeeding jobs (e.g. m_{i+1}, which leads to non-repeatable read/phantom read anomalies).

An example is depicted in Fig. 4. The CDC process is continuously running and sending captured deltas from OLTP sources (e.g. transaction log) to the ETL maintenance flow which propagates updates to warehouse tables on which OLAP queries are executed. In our example, the CDC process has successfully extracted delta tuples of three committed transactions T_1, T_2 and T_3 from the transaction log files and buffered them in the DPA of the ETL maintenance flows. The first query Q_1 occurs at the warehouse side at time t_2. The execution of Q_1 is first suspended until its relevant warehouse tables are updated by maintenance flows using available captured deltas of T_1 and T_2 which are committed before t_2. The delta tuples of T_1 and T_2 are grouped together as an input delta batch with the id of the maintenance job m_1. Once m_1 is finished, Q_1 is resumed and sees an up-to-date snapshot. The execution of the second query Q_2 (at t_3) forces the warehouse table state to be upgraded with another maintenance job m_2 with only source deltas derived from T_3. Note that, due to serializable snapshot isolation mechanism, the execution of Q_1 always uses the same snapshot that is taken from the warehouse tables refreshed with the final delta batch of m_1, and will not be affected by the new state that is demanded by Q_2. The third query Q_3 occurs at $t_{3,5}$ preceding the commit time of T_4. Therefore, no additional delta needs to be propagated for answering Q_3 and it shares the same snapshot with Q_2.

In our work, we assume that the CDC is always capable of delivering up-to-date changes to the DPA for real-time analytics. However, this assumption normally does not hold in reality and maintenance anomalies might occur in this situation as addressed by Zhuge et al. [5]. In Fig. 4, there is a CDC delay between the recording time of T_4's delta tuples in the transaction log and their occurrence time in the DPA of the ETL flow. The occurrence of the fourth query Q_4 arriving at t_4 requires a new warehouse state updated by the deltas of T_4 which are still not available in the DPA. We provide two realistic options here to compensate for current CDC implementations. The first option is to relax the query consistency of Q_4 and let it share the same snapshot with Q_2 and Q_3. The OLAP queries can tolerate small delays in updates and a "tolerance window" can be set (e.g., 30 s or 2 min) to allow scheduling the query without having to wait for all updates to arrive. This tolerance window could be set arbitrarily. Another option is to force maintenance processing to hang on until the CDC has successfully delivered all required changes to the DPA with known scope of input deltas for answering Q_4. With these two options, we continue with introducing our workload scheduler and incremental ETL pipeline based on the scope of our work depicted in Fig. 1.

5   Workload Scheduler

As we defined the consistency notion in the previous section, the suspended execution of any incoming query resumes only if relevant tables are refreshed by corresponding final delta batch. Updating warehouse tables is normally done by the last (sink) operator in our incremental ETL pipeline and transactions are run to permanently install updates from multiple delta batches into warehouse tables. We denote the transactions running in the last sink operator thread as *sink transactions* (ST). In this section, we focus on our workload scheduler which is used to orchestrate the execution of sink transactions and OLAP queries. Integrity constraints are introduced which deliver an execution order of begin and commit actions among sink transactions and OLAP queries.

Recall that an event of a query arrival Q_i immediately triggers the creation of a new maintenance job m_i, which updates the warehouse state for Q_i. The execution of Q_i is suspended until m_i is completed in the ST_i (i.e. the i-th transaction execution of ST commits successfully with its commit action $c(ST_i)$). Query Q_i is later executed in a transaction as well in which the begin action (denoted as $b(Q_i)$) takes a snapshot of the new warehouse state changed by ST_i. Therefore, the first integrity constraint enforced by our workload scheduler is $t(c(ST_i)) < t(b(Q_i))$ which means that ST_i should be committed before Q_i starts.

With arrivals of a sequence of queries $\{Q_i, Q_{i+1}, Q_{i+2}, ...\}$, a sequence of corresponding sink transactions $\{ST_i, ST_{i+1}, ST_{i+2}, ...\}$ are run for corresponding final delta batches. Note that, once the $b(Q_i)$ successfully happens, the query Q_i does not block its successive sink transaction ST_{i+1} for consistency control since the snapshot taken for Q_i is not interfered by ST_{i+1}. Hence, $\{ST_i, ST_{i+1}, ST_{i+2}, ...\}$ can run concurrently and commit in order while each $b(Q_i)$ is aligned with the end of its corresponding $c(ST_i)$ into $\{c(ST_i), b(Q_i), c(ST_{i+1}), ...\}$. However, only with the first constraint, the serializability property is still not guaranteed since the commit action $c(ST_{i+1})$ of a simultaneous sink transaction execution might precede the begin action $b(Q_i)$ of its preceding query. For example, after ST_i is committed, the following ST_{i+1} might be executed and committed so fast that Q_i has not yet issued the begin action. The snapshot now taken for Q_i includes rows updated by deltas occurring later than Q_i's submission time, which incurs non-repeatable/phantom read anomalies. In order to avoid these issues, the second integrity constraint is $t(b(Q_i)) < t(c(ST_{i+1}))$. This means that each sink transaction is not allowed to commit until its preceding query has successfully begun. Therefore, a serializable schedule can be achieved if the integrity constraint $t(c(ST_i)) < t(b(Q_i)) < t(c(ST_{i+1}))$ is not violated. The warehouse state is incrementally maintained by a sequence of consecutive sink transactions in response to the consistent snapshots required by incoming queries.

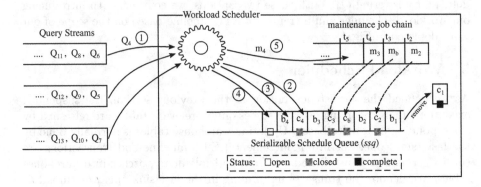

Fig. 5. Scheduling sink transactions and OLAP queries

Figure 5 illustrates the implementation of the workload scheduler. An internal queue called *ssq* is introduced for a serializable schedule of sink and query transactions. Each element e in *ssq* represents the status of a corresponding transaction and serves as a waiting point to suspend the execution of its transaction. We also introduced the three levels of query consistency (i.e. *open*, *closed* and *complete*) defined in [6] in our work to identify the status of the sink transaction. At any time there is always one and only one open element stored at the end of *ssq* to indicate an open sink transaction (which is ST_4 in this example). Once a query (e.g. Q_4) arrives at the workload scheduler ①, the workload scheduler first changes the status of the last element in *ssq* from open to closed. This indicates that the maintenance job for a pending ST_4 has been created and the commitment c_4 of ST_4 should wait on the completion of this *ssq* element ②. Furthermore, a new element b_4 is pushed into *ssq* which suspends the execution of Q_4 before its begin action ③. Importantly, another new open element is created and put at the end of *ssq* to indicate the status of a subsequent sink transaction triggered by the following incoming query (e.g. Q_5) ④. The ST_4 is triggered to be started afterwards ⑤. When ST_4 is done and all the deltas have arrived at warehouse site, it marks its *ssq* element c_4 with complete and keeps waiting until c_4 is removed from *ssq*. Our workload scheduler always checks the status of the head element of *ssq*. Once its status is changed from closed to complete, it removes the head element and notifies the corresponding suspended transaction to continue with subsequent actions. In this way, the commitment of ST_4 would never precede the beginning of Q_3 which takes a consistent snapshot maintained by its preceding maintenance transactions $\{ST_2, ST_b{}^3, ST_3\}$. Besides, Q_4 begins

---

[3] A system-level maintenance job is constructed and executed by the ST_b transaction, as certain source delta stream exceeds a pre-defined threshold.

only after ST_4 has been committed. Therefore, the constraints are satisfied and a serializable schedule is thereby achieved.

6 Operator Thread Synchronization and Coordination

In the Sect. 3, we see that the incremental ETL pipeline is capable of handling multiple maintenance jobs simultaneously. However, for those operator threads which read and write the same intermediate staging tables or warehouse dimension tables in the same pipeline, inconsistencies can still arise in the final delta batch. In this section, we first address inconsistency anomalies in two cases: incremental join and slowly changing dimensions. After that, we introduce a new concept of *consistency zone* which is used to synchronize/coordinate operator threads for consistent target deltas. In the end, we discuss the options to improve the efficiency of an incremental ETL pipeline with consistency zones.

6.1 Pipelined Incremental Join

An incremental join is a logical operator which takes the deltas (insertions, deletions and updates) on two join tables as inputs and calculates target deltas for previously derived join results. In [3], a delta rule[4] was defined for incremental joins (shown as follows). Insertions on table R are denoted as ΔR and deletions as ∇R. Given the old state of the two join tables (R_{old} and S_{old}) and corresponding insertions (ΔR and ΔS), new insertions affecting previous join results can be calculated by first identifying matching rows in the mutual join tables for the two insertion sets and further combining the newly incoming insertions found in ($\Delta R \bowtie \Delta S$). The same applies to detecting deletions.

$$\Delta(R \bowtie S) \equiv (\Delta R \bowtie S_{old}) \cup (R_{old} \bowtie \Delta S) \cup (\Delta R \bowtie \Delta S)$$

$$\nabla(R \bowtie S) \equiv (\nabla R \bowtie S_{old}) \cup (R_{old} \bowtie \nabla S) \cup (\nabla R \bowtie \nabla S)$$

For simplicity, we use the symbol Δ to denote all insertions, deletions and updates in this paper. Hence, the first rule is enough to represent incremental join with an additional join predicate (R.action = S.action) added to ($\Delta R \bowtie \Delta S$) where action can be insertion I, deletion D or update U.

We see that a logical incremental join operator is mapped to multiple physical operators, i.e. three join operators plus two union operators. To implement this delta rule in our incremental ETL pipeline, two tables R_{old} and S_{old} are materialized in the staging area during historical load and two extra update operators (denoted as $\uplus$) are introduced. One $\uplus$ is used to gradually maintain the staging table S_{old} using the deltas ($\Delta_{m_1}S, \Delta_{m_2}S, ...\Delta_{m_{i-1}}S$) from the executions of preceding maintenance jobs ($m_1, m_2, ..., m_{i-1}$) to bring the join table S_{old} to *consistent* state $S_{m_{i-1}}$ for $\Delta_{m_i}R$:

$$S_{m_{i-1}} = S_{old} \uplus \Delta_{m_1}S... \uplus \Delta_{m_{i-1}}S$$

---

[4] Updates are treated as deletions followed by insertions in this rule.

Another update operator $\uplus$ performs the same on the staging table R_{old} for $\Delta_{m_i}S$. Therefore, the original delta rule is extended in the following based on the concept of our maintenance job chain.

$$\Delta_{m_i}(R \bowtie S) \equiv (\Delta_{m_i}R \bowtie S_{m_{i-1}}) \cup (R_{m_{i-1}} \bowtie \Delta_{m_i}S) \cup (\Delta_{m_i}R \bowtie \Delta_{m_i}S)$$
$$\equiv (\Delta_{m_i}R \bowtie (S_{old} \uplus \Delta_{m_{1 \sim (i-1)}}S)) \cup ((R_{old} \uplus \Delta_{m_{1 \sim (i-1)}}R) \bowtie \Delta_{m_i}S)$$
$$\cup (\Delta_{m_i}R \bowtie \Delta_{m_i}S)$$

The deltas $\Delta_{m_i}(R \bowtie S)$ of job m_i are considered as *consistent* only if the update operators have completed job $m_{(i-1)}$ on two staging tables before they are accessed by the join operators. However, our ETL pipeline only ensures that the maintenance job chain is executed in sequence in each operator thread. Inconsistency can occur when directly deploying this extended delta rule in our ETL pipeline runtime. This is due to concurrent executions of join and update operators on the same staging table for different jobs.

Customer (refreshed by m_1)

id	name	company
1	bob	IBM
2	mary	SAP

Company (refreshed by m_1)

name	nation
IBM	USA

Δ(Customer $\bowtie$ Company)

job	action	value
m_2	—	$\varnothing$
m_3	I	(3, 'jack', 'HP', 'USA') (2, 'mary', 'SAP', 'Germany')
m_4	I	(4, 'peter', 'SAP', 'Germany')

ΔCustomer

job	action	value
m_2	—	—
m_3	I	(3, 'jack', 'HP')
m_4	I	(4, 'peter', 'SAP')

ΔCompany

job	action	value
m_2	I	('HP', 'USA')
m_3	I	('SAP', 'Germany')
m_4	—	—

Incorrect Δ(Customer $\bowtie$ Company):
$(\Delta_{m_3}\text{Customer} \bowtie \text{Company}_{m_1}) \cup (\text{Customer}_{m_1} \bowtie \Delta_{m_3}\text{Company})$
$\cup (\Delta_{m_3}\text{Customer} \bowtie \Delta_{m_3}\text{Company})$

job	action	value
m_3	I	(2, 'mary', 'SAP', 'Germany') (4, 'peter', 'SAP', 'Germany')

Fig. 6. Anomaly example for pipelined incremental join

We use a simple example (see Fig. 6) to explain the potential anomaly. The two staging tables Customer and Company are depicted at the left-upper part of Fig. 6 which both have been updated by deltas from m_1. Their input delta streams are shown at left-bottom part and each of them contains a list of tuples in the form of (job, action, value) which is used to store insertion-/deletion-/update-delta sets (only insertions with action I are considered here) for each maintenance job. Logically, by applying our extended delta rule, consistent deltas Δ(Customer $\bowtie$ Company) would be derived which are shown at the right-upper part. For job m_3, a matching row ('HP', 'USA') can be found from the company table for a new insertion (3, 'jack', 'HP') on the customer table after the company table was updated by the preceding job m_2. With another successful row-matching between Δ_{m_3}Company and Customer$_{m_2}$, the final deltas are complete and correct.

However, since at runtime, each operator thread runs independently and has different execution latencies for inputs of different sizes, an inconsistent case can occur which is shown at the right-bottom part. Due to various processing costs, the join operator Δ_{m_3} Customer $\bowtie$ Company $_{m_1}$ has already started before the update operator completes m_2 on the company table, which mistakenly missed the matching row ('HP', 'USA') from $m2$. And the other join operator Customer $_{m_4}$ $\bowtie$ Δ_{m_3} Company accidentally reads a phantom row (4, 'peter', 'SAP') from the maintenance job m_4 that is accomplished by the fast update operator on the customer table. This anomaly is caused by a pipeline execution without synchronization of read-/write-threads on the same staging table.

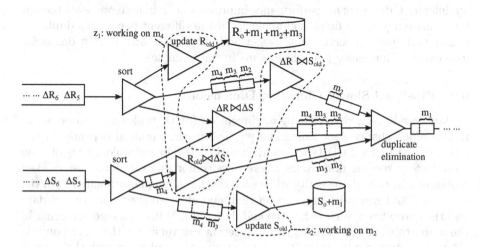

Fig. 7. Pipelined incremental join with consistency zones

To address this problem, we propose a *pipelined incremental join* for the maintenance job chain. It is supported by newly defined *consistency zones* and an extra duplicate elimination operator. Figure 7 shows the implementation of our pipelined incremental join[5]. In a consistency zone, operator thread executions are synchronized on the same maintenance job and processing of a new maintenance job is not started until all involving operator threads have completed the current one. This can be implemented by embedding a *cyclic barrier*(cb) (Java) object in all covered threads. Each time a new job starts in a consistency zone, this cb object sets a local count to the number of all involved threads. When a thread completes, it decrements the local count by one and blocks until the count becomes zero. In Fig. 7, there are two consistency zones: z_1(update-R_{old}, R_{old} $\bowtie$ ΔS) and z_2(Δ R $\bowtie$ S_{old}, update-S_{old}), which groups together all the threads that read and write the same staging table. The processing speeds of both threads in z_1 are very similar and fast, so both of them are currently working on

---

[5] The two sort operators are just required for merge join and can be omitted if other join implementations are used.

m_4 and there is no new maintenance job buffered in any of the in-memory pipes of them. However, even though the original execution latency of the join operator thread $\Delta R \bowtie S_{old}$ is low, it has to be synchronized with the slow operator update-S_{old} on m_2 and a pile-up of maintenance jobs (m_{2-4}) exists in its input pipe. It is worth to note that a strict execution sequence of two read-/write threads is not required in a consistency zone (i.e. update-R_{old} does not have to start only after $R_{old} \bowtie \Delta S$ completes to meet the consistency requirement $R_{m_{i-1}} \bowtie \Delta_{m_i} S$). In case $R_{m_{i-1}} \bowtie \Delta_{m_i} S$ reads a subset of deltas from m_i (in R) due to concurrent execution of update-$R_{m_{i-1}}$ on m_i, duplicates will be deleted from the results of $\Delta_{m_i} R \bowtie \Delta_{m_i} S$ by the downstream duplicate elimination operator. Without a strict execution sequence in consistency zones, involved threads can be scheduled on different CPU cores for performance improvement. Furthermore, even though two consistency zones finish maintenance jobs in different paces, this duplicate elimination operator serves as a *Merger* and only reads correct input deltas for its current maintenance job, which is m_2 in the example.

6.2   Pipelined Slowly Changing Dimensions

In data warehouses, slowly changing dimension (SCD) tables need to be maintained which change over time. The physical implementation depends on the type of SCD (three SCD types are defined in [9]). For example, SCDs of type 2 are history-keeping dimensions where rows comprising the same business key represent a history of one entity while each row has a unique surrogate key in the warehouse and was valid in a certain time period (from start date to end date and the current row version has the end date null). With a change occurring in the source table of a SCD table, the most recent row version of the corresponding entity (end date is null) is updated by replacing the null value with the current date and a new row version is inserted with a new surrogate key and a time range (current date - null). In the fact table maintenance flow, the surrogate key of this current row version of an entity is looked up as a foreign key value in the fact table.

Assume that the source tables that are used to maintain fact tables and SCDs reside in different databases. A globally serializable schedule S of the source actions on these source tables needs to be replayed in ETL flows for strong consistency in data warehouses [12]. Otherwise, a consistency anomaly can occur which will be explained in the following (see Fig. 8).

At the upper-left part of Fig. 8, two source tables: plin and item-S are used as inputs for a fact table maintenance flow (Flow 1) and a dimension maintenance flow (Flow 2) to refresh warehouse tables sales and item-I, respectively. Two source-local transactions T_1 (start time: $t_1 \sim$ commit time: t_2) and T_3 ($t_4 \sim t_6$) have been executed on *item-S* to update the price attribute of an item with business key ('abc') in one source database. Two additional transactions T_2 ($t_3 \sim t_5$) and T_4 ($t_7 \sim t_8$) have been also completed in a different database where a new state of source table *plin* is affected by two insertions sharing the same business key ('abc'). Strong consistency of the warehouse state can be reached if the globally serializable schedule S: $T_1 \leftarrow T_2 \leftarrow T_3 \leftarrow T_4$ is also guaranteed

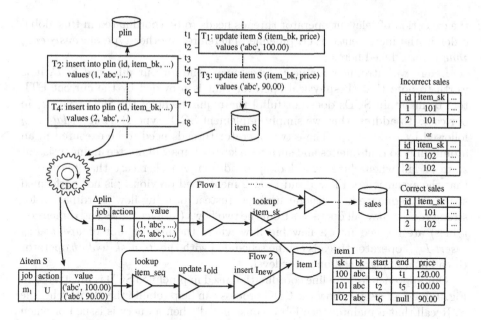

Fig. 8. Anomaly example for ETL pipeline execution without coordination

in ETL pipeline execution. A consistent warehouse state has been shown at the bottom-right part of Fig. 8. The surrogate key (101) found for the insertion (1, 'abc', ...) is affected by the source-local transaction T_1 on *item-S* while the subsequent insertion (2, 'abc', ...) will see a different surrogate key (102) due to T_3. However, the input delta streams only reflect the local schedules S_1: $T_1 \leftarrow T_3$ on *item-S* and S_2: $T_2 \leftarrow T_4$ on *plin*. Therefore, there is no guarantee that the global schedule S will be correctly replayed since operator threads run independently without coordination. For example, at time t_9, a warehouse query occurs, which triggers an immediate execution of a maintenance job m_1 that brackets T_2 and T_4 together on *plin* and groups T_1 and T_3 together on *item-S*. Two incorrect states of the *sales* fact table have been depicted at the upper-right part of the figure. The case where *item_sk* has value 101 twice corresponds to an incorrect schedule: $T_1 \leftarrow T_2 \leftarrow T_4 \leftarrow T_3$ while another case where *item_sk* has value 102 twice corresponds to another incorrect schedule: $T_1 \leftarrow T_3 \leftarrow T_2 \leftarrow T_4$. This anomaly is caused by an uncontrolled execution sequence of three read-/write-operator threads: *item_sk-lookup* in Flow 1 and *update-I_{old}* and *insert-I_{new}* in Flow 2.

To achieve a correct globally serializable schedule S, the CDC component should take the responsibility of rebuilding S by first tracking start or commit timestamps of source-local transactions[6], mapping them to global timestamps and finally comparing them to find out a global order of actions. In addition,

---

[6] Execution timestamps of in-transaction statements have to be considered as well, which is omitted here.

the execution of relevant operator threads needs to be coordinated in this global order in the incremental ETL pipeline. Therefore, another type of *consistency zone* is introduced here.

Before we introduce our new consistency zone for our *pipelined SCD*, it is worth to note that the physical operator that is provided by the current ETL tool to maintain SCDs does not fulfill the requirement of the SCD (type 2) in our case. To address this, we simply implement SCD (type 2) using *update-I_{old}* followed by *insert-I_{new}*. These two operator threads need to be executed in an atomic unit so that queries and surrogate key lookups will not see an inconsistent state or fail when checking a lookup condition. Another case that matters is that the execution of Flow 1 and Flow 2 mentioned previously is not performed strictly in sequence in a disjoint manner. Instead of using flow coordination for strong consistency, all operators from the two flows (for fact tables and dimension tables) are merged into a new big flow where the atomic unit of *update-I_{old} insert-I_{new}* operator threads can be scheduled with the *item_sk-lookup* operator thread at a fine-grained operator level.

Our approach for pipeline coordination used in pipelined SCD is illustrated in Fig. 9. We first explain how the CDC process can help rebuild the global schedule S. Recall that a maintenance job is constructed when a query is issued or when the size of any input delta stream exceeds a threshold (see Sect. 3). We refine the maintenance job into multiple internal, fine-grained *tasks* whose construction is triggered by a commit action of a source-local transaction affecting the source table of a SCD.

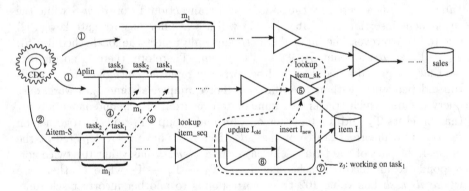

Fig. 9. Pipelined SCD with consistency zone

As shown in Fig. 9, ① the CDC continuously puts those captured source deltas into the input delta streams (one is $\Delta plin$) of the fact table maintenance flow. At this time, a source-local update transaction commits on *item-S*, which creates a $task_1$ and comprises the delta tuples derived from this update transaction ②. This immediately creates another $task_1$ in the input delta stream $\Delta plin$ which contains all current available delta tuples ③. This means that all source-local, update transactions belonging to the $task_1$ in $\Delta plin$ have committed before the $task_1$ of $\Delta item\text{-}S$. With a commit of the second update transaction on source table $item\text{-}S$, two new $task_2$ are created in both input delta streams ④. When a query is issued at a later time, a new m_1 is constructed which contains $task_{1\sim2}$ on $\Delta item\text{-}S$ and $task_{1\sim3}$ on $\Delta plin$ (delta tuples in $task_3$ commit after the $task_2$ in $\Delta item\text{-}S$). During execution on m_1, a strict execution sequence between the atomic unit of $update\text{-}I_{old}$ and $insert\text{-}I_{new}$ and the $item\_sk\text{-}lookup$ is forced for each $task_i$ in m_1. The $update\text{-}I_{old}$ and $insert\text{-}I_{new}$ have to wait until the $item\_sk\text{-}lookup$ finishes $task_1$ ⑤ and the $item\_sk\text{-}lookup$ cannot start to process $task_2$ until the atomic unit completes $task_1$ ⑥. This strict execution sequence can be implemented by the (Java) *wait/notify* methods as a provider-consumer relationship. Furthermore, in order to guarantee the atomic execution of both $update\text{-}I_{old}$ and $insert\text{-}I_{new}$ at task level, (Java) *cyclic barrier* can be reused here to let $update\text{-}I_{old}$ wait to start a new task until $insert\text{-}I_{new}$ completes the current one ⑥. Both thread synchronization and coordination are covered in this consistency zone ⑦.

6.3   Discussion

In several research efforts on operator scheduling, efficiency improvements can be achieved by cutting a data flow into several sub-flows. In [14], one kind of sub-flow called *superboxes* are used to batch operators to reduce the scheduling overhead. And authors of [15] use another kind of sub-flow (*strata*) to exploit pipeline parallelism to some extent. In this work, the operators involved in a sub-flow are normally connected through data paths. However, as described in the previous two sections, consistency zones can have operator threads scheduled together without any connecting data path. This increases the complexity of algorithms which try to increase the pipeline efficiency as much as possible by minimizing the execution time of operator with $max(time(op_i))$. However, we will not validate the performance of the scheduling algorithms extended for consistency zones using experiments in this paper. A pipeline that was previously efficient can be slowed down dramatically when one of its operator is bound with a very slow operator in a consistency zone, which increases the $max(time(op_i))$.

The efficiency of an incremental ETL pipeline with consistency zones could be improved if the data storage supports multi-version concurrency control, where reads do not block writes and vice versa. Therefore, a fast update operator on a staging table will not be blocked by a slow join operator which reads rows using version number (possibly maintenance job id in our case). However, in another case, a fast join operator may still have to wait until the deltas with the desired version are made available by a slow update operator.

7   Experimental Results

We examine the performance in this section with read-/update-heavy workloads running on three kinds of configuration settings.

Test Setup: We used the TPC-DS benchmark [11] in our experiments. Our testbed is composed of a target warehouse table *store sales* (SF 10) stored in a Postgresql (version 9.4) instance which was fine-tuned, set to serializable isolation level and ran on a remote machine (2 Quad-Core Intel Xeon Processor E5335, 4 × 2.00 GHz, 8 GB RAM, 1 TB SATA-II disk), an ETL pipeline (an integrated pipeline instance used to update item and store sales tables) running locally (Intel Core i7-4600U Processor, 2 × 2.10 GHz, 12 GB RAM, 500 GB SATA-II disk) on an extended version of Kettle (version 4.4.3) together with our workload scheduler and a set of query streams, each of which issues queries towards the remote store sales table once at a time. The maintenance flow is continuously fed by deltas streams from a CDC thread running on the same node. The impact of two realistic CDC options (see Sect. 4) was out of scope and not examined.

We first defined three configuration settings as follows.

Near Real-time (NRT): simulates a general near real-time ETL scenario where only one maintenance job was performed concurrently with query streams in a small time window. In this case, there is no synchronization of maintenance flow and queries. Any query can be immediately executed once it arrives and the consistency is not guaranteed.

PipeKettle: uses our workload scheduler to schedule the execution sequence of a set of maintenance transactions and their corresponding queries. The consistency is thereby ensured for each query. Furthermore, maintenance transactions are executed using our incremental ETL pipeline.

Sequential execution (SEQ): is similar to **PipeKettle** while the maintenance transactions are executed sequentially using a flow instance once at a time.

Orthogonal to these three settings, we simulated two kinds of read-/update-heavy workloads in the following.

Read-heavy workload: uses one update stream (SF 10) consisting of *purchases* (♯: 10 K) and *lineitems* (♯: 120 K) to refresh the target warehouse table using the maintenance flow and meanwhile issues totally 210 queries from 21 streams, each of which has different permutations of 10 distinct queries (generated from 10 TPC-DS ad-hoc query templates, e.g. q[88]). For PipeKettle and SEQ, each maintenance job consists of 48 new purchases and 570 new lineitems in average.

Update-heavy workloads: uses two update streams (♯: 20 K & 240 K) while the number of query streams is reduced to seven (totally 70 queries). Before executing a query in PipeKettle and SEQ, the number of deltas to be processed is 6-times larger than that in read-heavy workloads.

Test Results: Figure 10 illustrates a primary comparison among NRT, PipeKettle and SEQ in terms of flow execution latency without query interventions. As the baseline, it took 370 s for NRT to processing one update stream. The update stream was later split into 210 parts as deltas batches for PipeKettle and SEQ. It can be seen that the overall execution latency of processing 210 maintenance jobs in PipeKettle is 399 s which is nearly close to the baseline due to pipelining

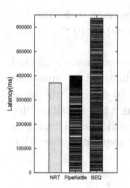

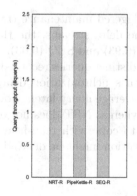

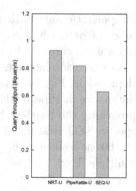

Fig. 10. Performance comparison without queries

Fig. 11. Query throughput in read-heavy workload

Fig. 12. Query throughput in Update-heavy workload

parallelism. However, the same number of maintenance jobs is processed longer in SEQ (∼650 s, which is significantly higher than the others).

Figure 11 and 12 show the query throughputs measured in three settings using both read-/update-heavy workloads. Since the maintenance job size is small in read-heavy workload, the synchronization delay for answer each query is also small. Therefore, the query throughput achieved by PipeKettle (2.22 queries/s) is very close to the one in baseline NRT (2.30) and much higher than the sequential execution mode (1.37). We prove that our incremental pipeline is able to achieve high query throughput at a very high query rate. However, in update-heavy workload, the delta input size becomes larger and the synchronization delay grows increasingly, thus decreasing the query throughput in PipeKettle. Since

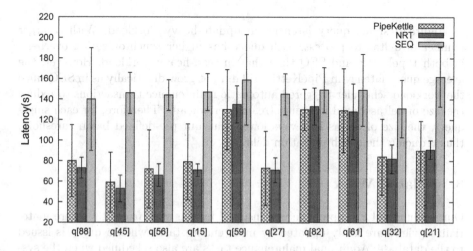

Fig. 13. Average latencies of 10 ad-hoc query types in read-heavy workload

our PipeKettle automatically triggered maintenance transactions to reduce the number of deltas buffered in the delta streams, the throughput (0.82) is still acceptable as compared to NRT(0.93) and SEQ (0.63).

The execution latencies of 10 distinct queries recorded in read-heavy workload is depicted in Fig. 13. Even with synchronization delay incurred by snapshot maintenance in PipeKettle, the average query latency over 10 distinct queries is approaching the baseline NRT whereas NRT does not ensure the serializability property. SEQ is still not able to cope with read-heavy workload in terms of query latency, since a query execution might be delayed by sequential execution of multiple flows.

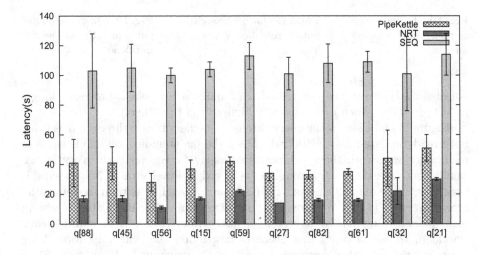

Fig. 14. Average latencies of 10 ad-hoc query types in update-heavy workload

Figure 14 shows query latencies in update-heavy workload. With a larger number of deltas to process, each query has higher synchronization overhead in both PipeKettle and SEQ than that in read-heavy workload. However, the average query latency in PipeKettle still did not grow drastically as in SEQ since the workload scheduler triggered automatic maintenance transactions to reduce the size of deltas stored in input streams periodically. Therefore, for each single query, the size of deltas is always lower than our pre-defined batch threshold, thus reducing the synchronization delay.

8   Related Work

Our incremental ETL pipeline was inspired by the work from [13] where materialized views are lazily updated by maintenance tasks when a query is issued to the database. Additional maintenance tasks are also scheduled when the system has free cycles to hide maintenance overhead partially from query response time. As we mentioned in the introduction section, ETL flows can be seen as a counterpart to the view definitions in databases where materialized view maintenance is performed in transactions to ensure the consistency property. Therefore, we addressed the potential consistency anomalies in general ETL engines which normally lack global transaction support.

Thomsen et al. addressed on-demand, fast data availability in so-called right-time DWs [4]. Rows are buffered at the ETL producer side and flushed to DW-side, in-memory tables with bulk-load insert speeds using different (e.g. immediate, deferred, periodic) polices. A timestamp-based approach was used to ensure accuracy of read data while in our work we used an internal queue to schedule the workloads for our consistency model. Besides, we also focused on improving throughput by extending an ETL tool.

Near real-time data warehousing was previously referred to as active data warehousing [2]. Generally, incremental ETL flows are executed concurrently with OLAP queries in a small time window. In [1], Vassiliadis et al. detailed a uniform architecture and infrastructure for near real-time ETL. Furthermore, in [3], performance optimization of incremental recomputations was addressed in near real-time data warehousing. In our experiments, we compare general near real-time ETL approach with our work which additionally guarantees the query consistency.

In [6,7], Golab et al. proposed temporal consistency in a real-time stream warehouse. In a certain time window, three levels of query consistency regarding a certain data partition in warehouse are defined, i.e. open, closed and complete, each which becomes gradually stronger. As defined, the status of a data partition is referred to open for a query if data exist or might exist in it. A partition at the level of closed means that the scope of updates to partition has been fixed even though they haven't arrived completely. The strongest level complete contains closed and meanwhile all expected data have arrived. We leverage these definitions of temporal consistency levels in our work.

9   Conclusion

In this work, we addressed the on-demand snapshot maintenance policy in MVCC-supported data warehouse systems using our incremental ETL pipeline. Warehouse tables are refreshed by continuous delta batches in a query-driven manner. We discussed a logical computational model and described the incremental ETL pipeline as a runtime implementation which addresses the performance challenge. Moreover, based on the consistency model defined in this paper, we introduced the workload scheduler which is able to achieve a serializable schedule of concurrent maintenance flows and OLAP queries. We extended an open-source ETL tool (Kettle) as the platform of running our incremental ETL pipeline and also addressed potential inconsistency anomalies in the cases of incremental join and slowly changing dimension tables by proposing the consistency zone concept. The experimental results show that our approach achieves average performance which is very close to traditional near real-time ETL while the query consistency is still guaranteed.

References

1. Vassiliadis, P., Simitsis, A.: Near real time ETL. In: Kozielski, S., Wrembel, R. (eds.) New Trends in Data Warehousing and Data Analysis, pp. 1–31. Springer, Boston (2009)
2. Karakasidis, A., Vassiliadis, P., Pitoura, E.: ETL queues for active data warehousing. In: Proceedings of the 2nd International Workshop on Information Quality in Information Systems, pp. 28–39. ACM (2005)
3. Behrend, A., Jörg, T.: Optimized incremental ETL jobs for maintaining data warehouses. In: Proceedings of the Fourteenth International Database Engineering & Applications Symposium, pp. 216–224. ACM (2010)
4. Thomsen, C., Pedersen, T.B., Lehner, W.: RiTE: providing on-demand data for right-time data warehousing. In: ICDE, pp. 456–465 (2008)
5. Zhuge, Y., Garcia-Molina, H., Hammer, J., Widom, J.: View maintenance in a warehousing environment. ACM SIGMOD Rec. $24(2)$, 316–327 (1995)
6. Golab, L., Johnson, T.: Consistency in a stream warehouse. In: CIDR. Vol. 11, pp. 114–122 (2011)
7. Golab, L., Johnson, T., Shkapenyuk, V.: Scheduling updates in a real-time stream warehouse. In: ICDE, pp. 1207–1210 (2009)
8. Kemper, A., Neumann, T.: HyPer: a hybrid OLTP&OLAP main memory database system based on virtual memory snapshots. In: ICDE, pp. 195–206 (2011)
9. Kimball, R., Caserta, J.: The Data Warehouse ETL Toolkit. Wiley, Hoboken (2004)
10. Casters, M., Bouman, R., Van Dongen, J.: Pentaho Kettle Solutions: Building Open Source ETL Solutions with Pentaho Data Integration. Wiley, Indianapolis (2010)
11. http://www.tpc.org/tpcds/
12. Zhuge, Y., Garcia-Molina, H., Wiener, J.L.: The Strobe algorithms for multi-source warehouse consistency. In: Fourth International Conference on Parallel and Distributed Information Systems, 1996, pp. 146–157. IEEE, December 1996
13. Zhou, J., Larson, P.A., Elmongui, H.G.: Lazy maintenance of materialized views. In: Proceedings of the 33rd International Conference on Very Large Data Bases, pp. 231–242. VLDB Endowment, September 2007
14. Carney, D., Çetintemel, U., Rasin, A., Zdonik, S., Cherniack, M., Stonebraker, M.: Operator scheduling in a data stream manager. In: Proceedings of the 29th International Conference on Very Large Data Bases, vol. 29, pp. 838–849. VLDB Endowment, September 2003
15. Karagiannis, A., Vassiliadis, P., Simitsis, A.: Scheduling strategies for efficient ETL execution. Inf. Syst. $38(6)$, 927–945 (2013)
16. Qu, W., Basavaraj, V., Shankar, S., Dessloch, S.: Real-time snapshot maintenance with incremental ETL pipelines in data warehouses. In: Madria, S., Hara, T. (eds.) DaWaK 2015. LNCS, vol. 9263, pp. 217–228. Springer, Cham (2015). doi:10.1007/978-3-319-22729-0_17

Author Index

Printed in the United States
By Bookmasters

Printed in the United States
By Bookmasters